SURFACE MOLECULAR CHEMISTRY

SURFACE MOLECULAR CHEMISTRY

Dr. Sonia Chahar Srivastava

Oxford Book Company

Jaipur ▪ New Delhi

Published By

Oxford Book Company

Regd. Off. "Bony Residency", Gate No. 2
Opp. Tilak Public School, Vishweswariya Nagar,
Gopalpura Road, Jaipur - 302018, India
Ph: 0141 - 2761280, Telefax: 2761381
email - oxfordbookjpr@gmail.com
website - oxfordbookcompany.com

Delhi Office: 102, 1st Floor, Satyam House,
4327/3, Ansari Road, Darya Ganj,
New Delhi - 110002, India
Ph: 011-45652440

Distributed by
KEYA INTERNATIONAL PRESS LLP
E-mail: keyainternationalpressllp@gmail.com

First published in India, 2024

ISBN : 9789355241153

Typeset by Shivangi Computers, Jaipur

Printed and bound in India by Replika Press Pvt. Ltd.

Contents

Preface

The goal of the Surface Chemical Dynamics Program is to elucidate the underlying physical processes that determine the products (selectivity) and yield (efficiency) of chemical transformations relevant to energy-related chemistry on catalytic and nanostructured surfaces. Achieving this end requires understanding the evolution of the reactant-molecule/surface complex as molecules adsorb, bonds dissociate, surface species diffuse, new bonds form and products desorb. The pathways and time scales of these processes are ultimately determined by a multidimensional potential energy surface that is a function of the geometric and electronic structures of the surface and the reactant, product, intermediate and transition-state molecular and atomic species.

To understand these processes, our group employs a battery of methods to measure the energetics and timescales of energy and charge transfer in the molecule/surface complex. Excitation of the complex is brought about either by preparing the reactants in a molecular beam or exciting adsorbates with photons. The response of the system is probed by measuring sticking coefficients, rovibrational-state-, angle-, and time-resolved molecular desorption distributions, electronic energy distribution dynamics, photoinduced desorption dynamics and adsorbate vibrational energy relaxation dynamics. In addition we are developing surface nanostructuring methods, including physical vapour and size-selected cluster deposition that can be coupled to these powerful surface chemical dynamics probes. By exercising temporal control over the excitation and spatial control over the surface we aim to gain detailed insight into the dynamical links between the structure and function of reactive surfaces.

A molecule may scatter off the surface, experiencing no or some finite degree of energy exchange with the surface. Alternatively, molecule-surface energy transfer can lead to accommodation and physical adsorption or chemical adsorption. In some cases, physisorption is a precursor to chemisorption, and in some cases, bond dissociation is required for chemisorption. Charge transfer plays a critical role in some adsorption processes. Once on the surface, the adsorbed intermediate may diffuse laterally with a temperature dependent rate, sampling surface features including adatoms, vacancies and steps. They may become tightly bound to a defect site. Various adsorbed intermediates may meet, either at defect sites or at regular lattice sites, and form short-lived transition state structures and ultimately product molecules. Finally, products desorb from the surface with a temperature dependent rate, imparting some fraction of the energy of the association reaction to the surface. The goal of surface chemical dynamics is to identify the roles, quantify the rates and understand the physics of these various mechanistic steps that comprise the surface chemical transformation.

This book contains a brief review of absorption and catalysis, for practical purposes. This also presents some important information on adsorbents and catalysts.

— Dr. Sonia Chahar Srivastava

Chapter 1

Geochemistry of Sedimentary

The Iberian Pyrite Belt is the world's greatest concentration of large massive sulphide deposits associated with volcanic rocks and hence the palacovolcanological setting has to be well documented if the ore-forming process is to be understood. Currently no consensus exists on the volcanological style of the Volcano-Sedimentary Complex host to this important accumulation of metals and two contrasting models arc being debated. Since the adoption of the exhalative model for the mineralisation, the host succession has been interpreted as a dominantly effusive, and explosively eruptive, volcanic pile. More recently this model has been challenged by the emergence of the sill-sediment-complex model in which up to 90% of the host rocks are high-level pcperitic intrusions. The two volcanological models lead to very different conclusions about the chronology of the host rocks, the timing and genesis of alteration, the driving force for the ore-related hydrothermal system, the source of metals, and synmineralisation reconstructions.

A distinctive feature of the Volcano-Sedimentary Complex is the alternation of primary volcanic facies and sedimentary facies. This presents an opportunity to discriminate between the competing volcanogenic models because some provenance patterns in the sedimentary intervals are expected to be very different in the two settings. Saez et al. (1996) expressed the commonly held view of the volcanology when they noted 'a predominance of sub-aerial or shallow-water volcanic activity' and claimed strong support for 'a prevailing explosive mechanism for the felsic volcanic activity'. Such a volcanic style would be expected to contribute much volcanic detritus to sedimentary intervals. In sill-sediment complexes there does not have to be any genetic link between the sills and overlying sedimentary rocks. In the extreme end-member case, represented by the still evolving complex in the Guaymas Basin, none of the igneous rocks have reached the surface and, because they are totally confined, cannot provide detritus to the sedimentary environment. Sediments between the mafic sills in the Guaymas Basin are a mixture of continent-derived gravity-flow deposits and oceanic pelagic deposits. The other end of the spectrum in sill-sediment complexes is represented by cryptodomes that have evolved to become largely emergent where much of the overlying sediment is a carapace breccia, derived from the locally exposed igneous body. Between these end-members lies what is probably the most common sill-sediment situation where the cover to some intrusive sheets breaks down and hydrovolcanic eruptions feed fluidized mixtures of fragments of the igneous rock, and the host sediments, into the depositional environment. Intrusive sheets may, therefore, intrude entirely non-volcanogenic sediments and/or the products of earlier hydrovolcanism. In this latter case sills may or may not have chemistry comparable with that of the overlying sediments.

This study documents the geochemistry, and interprets the provenance, of sedimentary rocks within the Pyrite Belt volcanogenie host sequence and compares these data with predictions made by the two competing volcanic models about the expected nature of sedimentary rocks interlayered with primary volcanic facies.

Tectonic Setting of the Iberian Pyrite Belt

The Iberian Pyrite Belt forms part of the South Portuguese Zone, a tcrrane separated from the Iberian Massif by the exotic, oceanic PuIo do Lobo ophiolitic terrane. The Volcano-Sedimentary Complex is a volcanogenic sequence that averages less than 1 km in thickness and is set in an otherwise terrigenous succession. Rhyolites and rhyodacites are volumetrically dominant in the Volcano-Sedimentary Complex but its overall character is bimodal with a second mode of basaltic rocks; intermediate compositions are widespread but minor. The underlying Phyllite-Quartzite Group clastic sequence (sandy turbidites and black mudrocks) formed in the Devonian at a passive margin.

For the metallogcny of the region, the key tectonic event was the transition, late in Devonian times, from this drift phase of passive margin evolution to active extensional tectonics. This switch was induced by oblique convergence as the South Portuguese Zone was sutured to the main Iberian Massif. Most workers have interpreted the Volcano-Sedimentary Complex magmatism as being a consequence of lithosphcric stretching during the oblique suturing event.

Towards the end of the Viscan, as the collision progressed, a switch from transtension to transpression resulted in deformation propagating southwestwards. Ahead of this deformation front there is a record of synorogenic sedimentation (flysch) that is referred to as the Culm. Regional metamorphism associated with the orogeny varies from very low to low grade, generally in a north-south gradient from zeolite facies in the south of the South Portuguese Zone to upper greenschist facies in the north all mudrocks in the region are now slates, but sedimentary facies can be safely identified and, throughout the discussion, terms relating to pre-metamorphic state will be used. Strongly sericitized volcanic rocks are also well cleaved but many macro-textures can still be recognised.

Contrasting Volcanological Models for the Volcano Sedimentary Complex: Rio Tinto District

Because the Rio Tinto district has been documented in detail in the context of both volcanological models, this region provides the best opportunity to determine which interpretation gives the most satisfactory explanation of the observed sedimentary provenance relationships.

According to the volcanic-pile model, the Volcano-Sedimentary Complex around Rio Tinto has a simple tripartite lithostratigraphy that is laterally continuous over tens of kilometres. This comprises a 400-500 m thick Basic Complex at its base, succeeded by a 50-800 m thick Acid Complex, and an uppermost Transition Scries, which can be several hundred metres thick but is commonly a few tens of metres. In this interpretation, the Basic Complex consists of basalts, pillowed in places, and derived volcaniclastic rock together with background mudrock. The Acid Complex has been described by Garcia as consisting of 'very homogeneous acid volcanics (dacitic to rhyolitic in composition), with predominant pyroclastics, and locally abundant breccias, autobrcccias, and lavas', and he described the Transition Series as being a 'heterogeneous pyroclastic level' comprising 'polygcnetic breccias, litho-crystalline tuffs, chloritized-haematizcd-scricitizcd tuffs, and massive sulphides'.

The following section summarises the findings of Boulter et al. (2001). In their sill-sediment-complex model, the primary volcanic facics are interpreted as predominantly subconcordant sills and more than 80% of the Volcano-Sedimentary Complex is a bimodal suite of post-mineralisation felsic and mafic pepcritic intrusions. It is proposed that this fold structure was the site of extensional growth faulting, which marked the end of Phyllite-Quartzite Group style sedimentation and controlled the formation of a wedge-shaped unit of debris-flow deposits. Earlier workers referred to this unit as the 'Slate Conglomerate', a distinctive interval of thickly bedded and poorly organised pebbly mudrocks and black mudrocks that ends in several metres of black mud. Apart from a few per cent of andesitic pebbles, the great bulk of the pebbles are muddy sandstones, muddy siltstoncs and anoxic mudrocks; this unit was overwhelmingly sourced in continental regions.

Towards the end of the phase of extensional tectonics, magmas utilised the extensional faults and were emplaccd in the vicinity of these structures mainly as composite basaltic andesite to rhyolite high-level intrusions of hundreds of metres thickness; these magmas were largely emplaced directly beneath the 'Slate Conglomerate' and its immediate lateral equivalents. The andesitic component became locally eruptive as hydrovolcanic eruptions that were generated where the cover to peperitic sills was breached. These eruptions deposited pebble-grade volcaniclastic breccias of an intermediate composition within a few kilometres of the Rio Tinto Anticline. At greater distances these breccias pass laterally into andesitic turbiditcs a few tens of metres thick.

The major hydrothermal event that created the Rio Tinto massive sulphide sheet was initiated during the extensional faulting phase, i.e. syn-'Slate Conglomerate', but peaked after the basaltic andesite-rhyolite magmatism that is pervaded by the stockwork feeder to the exhalative mineralisation. Waning hydrothermal flow resulted in chert deposition over some of the sulphide sheet. For the Rio Tinto district, the mineralisation was followed by a pronounced change in the redox state of the stratified sequence. Sedimentation characterised by radiolarian-bearing red mudrocks is typical of the interval between the sulphide layer and the flysch of the Culm. Intercalated with the red mudrocks are felsic volcaniclastic rocks in which very thickly bedded gravel-grade breccias are much more important than sandy facies. Most of these are debris-flow deposits directly or indirectly derived from hydrovolcanic eruptions associated with felsic peperitic intrusions, only one of which is preserved in the region. This intrusion, on the eastern edge, is a microphyric quartz-feldspar porphyry of 100 m thickness and 400 m strike length. The post-mineralisation red mudrock unit contains one of the greatest lateral facies contrasts in the region. The felsic volcaniclastic rocks overlying the sulphide sheets are thinly bedded (millimetres to tens of millimetres thick) sandy turbidites in a mudrock-dominated sequence. Where there are no sulphide sheets, felsic gravel-grade deposits predominate, and beds metres to tens of metres thick are common. This distribution defines an antithetic relationship between breccia facies in the red mudrock sequences and stratiform mineralisation.

Until late in the evolution of the marine red-bed sequence, the only primary volcanic facies in the region were the felsic microphyric intrusion in the red mudrock unit and the premineralisation intermediate-felsic igneous rocks in the Rio Tinto Anticline. Subsequently, a bimodal suite, which represents more than 80% of the volume of the Volcano-Sedimentary Complex in the area shown, intruded this assemblage. Outside the Rio Tinto Anticline (i.e. in non-mineralised regions) the bimodal suite intruded the 20-30 m of stratified andcsitic volcaniclastic rocks, interbedded with black mudrocks, and the overlying hectomctric red mudrock assemblage, as well as locally cutting down into the Phyllitc-Quartzitc Group. The typical section comprises one or two subconcordant felsic sheets of several hundred metres thickness and 5-10 mafic sheets each of which is typically several tens of metres thick. The bimodal package of felsic and mafic sills shows regionally discordant relationships. Westwards both the felsic and mafic sills reach progressively higher levels in the stratigraphy defined by the sedimentary rocks. This trend continues until in the Odiel River the mafic sills intrude the red mudrock-felsic volcaniclastic facies association. At this western locality, immediately below the main felsic sheet, in 420 m of stratigraphy, two main dolerite sheets averaging 135 m thick and a 20 m thick dolerite sheet alternate with red mudrocks and cobble-grade felsic volcanic-derived breccias.

Reference Measured Section

An almost fully exposed section through the Volcano-Sedimentary Complex in the Jarama River provides a reference section for discussion of the provenance relationships. The data presented here mainly come from, or can be reliably correlated with, the logged section and the mapping of Oswin (1986) allows this section to be tied. To emphasize the characteristics of the bedded rocks, the right- and left-hand columns exclude the primary volcanic facies. For the volcanic-pile model, the central column would reflect the chronology of the volcanic accumulation. For the sill-sediment-complex model, the marginal columns, if stacked, represent the sequence of stratified rocks before the intrusion of the bimodal suite.

Primary Volcanic Fades

Below the lowest sheet of largely coherent mafic igneous rocks is the Phyllite-Quartzite Group, a sequence of turbidites and black mudrocks. The lowest third of the Volcano-Sedimentary Complex comprises mafic sheets, averaging 30 m thick, intercalated with sedimentary rocks. This zone is equivalent to the Basic Complex of Garcia (1980, 1990). Interleaved with the mafic sheets are turbiditic sedimentary rocks in a background of black mudrock.

Petrographic and geochemical data on the mafic sheets support the field classification. Where plotted on a Zr/ TiO_2 v. Nb/Y diagram they fall in the basaltic andesite and subalkaline basalt fields. The mafic rocks have the most extreme Ti/Nb ratios of all the igneous rocks in the district, being in excess of 980:1. Petrographically the mafic sheets are dominantly pyroxene and plagioclase in near equal proportions. The feldspars are randomly oriented and subophitic textures arc common. Grain sizes are commonly 0.5-3 mm.

The three thick felsic units in the logged section are distinctive because of their lack of stratification; there are only 10 m of sedimentary rock in these 300 m. This felsic interval is equivalent to the Acid Complex of Garcia (1980, 1990). The felsic sheets in the Jarama River section are quartz-feldspar porphyries and they all plot in the rhyodacite-dacitc field of the Zr/TiO_2 v. Nb/Y diagram. In the Jarama section, all of the components of the felsic sheets have Ti/Nb ratios less than 400:1, and most are less than 300:1.

Sedimentary Fades

In the black mudrock interval, most of the sandy turbidites are medium sand grade and are mainly composed of T_a and T_b divisions of low-density turbidites; other divisions are rare. For the sandy beds logged in this portion of the Jarama section, the mean bed thickness is 4.96 cm, ranging from 2 mm for a fine-sand layer to 44 cm for a turbidite that has granules at its base.

Modal analysis of the framework grains has been carried out on the least deformed samples. The most striking feature of the volcanic turbidites associated with black mudrock is the virtual absence of quartz. The framework population of sample 8 is lithic clasts and heavily altered feldspar. More than 80% of the framework of sample 9 is lithic clasts, all similar in character. The rock fragments are mainly highly feldspar-phyric (quartz absent) with feldspar lath concentrations up to 50% set in a trachytic-textured groundmass. Rare aphyric clasts have, as their only recognisable grains, crowds of strongly aligned feldspar microlites c. 0.02 mm long. Above sample 9, in the uppermost final 5-10 m of black mudrock stratigraphy, lithic wackes contain phenocryst-rich clasts that have a minor (<2%) quartz-phyric component. Vesiculation of the clasts was very limited and clasts are typically bound by planar to slightly curviplanar fractures.

Above the felsic sheets in the Jarama section, there is a section of 70 m thickness dominated by felsic volcanic-derived gravelgrade volcaniclastic deposits where mudrocks are red (left-hand column, Fig. 4). This unit correlates with the lower part of the Transition Series of Garcia (1980, 1990). This gravel-dominated interval is largely thickly bedded, poorly organised sedimentary breccias. These beds commonly range from 5 to 15 m thick although the maximum is 32 m. These depositional units are composites of crudely graded sedimentary breccias at their bases and laminated pebbly sand facies in their upper portions. The boundary between the graded and laminated parts of the beds is locally marked by the presence of outsized clasts several tens of centimetres across compared with the centimetre-sized clasts below. Such a concentration of outsized clasts is believed to occur at the boundary between a lower traction carpet, which creates a debris-flow deposit, and an overlying turbulent cloud, which creates the laminated facies. It is possible that flow transformation within the body of the debris flows could have led to the development of turbulence, resulting in the uppermost laminated facies of these beds. The laminated facies commonly occupies around 20% of a bed but, in some cases, is found independently of the graded breccias.

There is considerable textural variation in the graded breccias, many of which consist of angular to subangular cobbles with rare outsize clasts up to several metres in diametre. Most beds are almost monomictic, comprising quartz-feldspar porphyry clasts, whereas some other beds are highly heterogeneous. The most admixed clast populations include felsic rocks with various phenocryst populations (quartz and/or feldspar in varying size and abundance), and sedimentary clasts such as red mudrock, chert and manganiferous chert. Clasts arc very rarely amygdaloidal and the generally dense (non-vesicular) character of the igneous clasts is a distinctive feature. Several per cent of the clasts are fragments of peperite.

Sedimentary Geochemistry

Major elements and selected trace elements were determined by XRF, following standard procedures, on a Phillips PW1400 instrument at the Southampton Oceanography Centre, using a 3 kW Rh-anode tube. Analytical precision is better than 5% (2 SD) for the major elements and 5% (2 SD) for the trace elements. Precision and accuracy are monitored using international standards analysed as unknowns. The REE were analysed by inductively coupled plasma mass spectrometry (ICP-MS) at the Southampton Oceanography Centre on a VG Plasmaquad II instrument. Precision is better than 4% (2 SD). In each sample, run precision was determined by using the international rock standard BHVO as the drift monitor after every five unknowns. A series of international rock standards was used to determine accuracy.

Petrogenetic studies on the igneous rocks of the Pyrite Belt have documented and discussed the nature of the non-stratified rocks. Here we consider the sedimentary rocks in the Volcano-Sedimentary Complex for which previous geochemical studies have tended to concentrate on the non-clastic components in the sequences that host the manganiferous deposits. The sedimentary geochemical data presented here have been organised into two main stratigraphie groupings according to the colour of their associated mudrock facies: black (lower) and red (upper). Within these, further grouping is based on sedimentary facies; namely, mudrock facies, sandy turbidites and gravel-grade sedimentary breccias.

The major- and trace-element chemistry for the stratified rocks is complex, as is shown by the lack of a single linear array on two-component Harker-type diagrams. Overall major-oxide concentrations show large ranges in weight per cent (SiO_2 92.75-59.55; Al_2O_3 20.28-2.75; Fe_2O_3 25.71-2.03; MgO 5.89-0.5; CaO 10.24-0.03; Na_2O 6.37-0.15; K_2O 5.71-0.24). However, there is a greater degree of consistency in the major oxide contents within the facies categories (black mudrock, red mudrock, sandy turbidites and sedimentary breccias associated with black mudrocks, and sandy turbidites and sedimentary breccias associated with red mudrocks). For example, mean MgO and CaO concentrations in the sandy turbidites and sedimentary breccias associated with the black mudrocks are around twice the level of those in sand- or gravelgrade sedimentary rocks associated with the red mudrocks. The values for Na_2O and K_2O are comparable for the coarse volcaniclastic rocks in both the red and black mudrock portions of the stratigraphy.

Overall concentrations of the ferromagnesian element Ni are very low, averaging 25 ppm. Although the concentrations of most high field strength elements are not particularly high (e.g. Nb average is 11 ppm, Th average 7 ppm), Zr in contrast, in the coarser of the stratified rocks intercalated with the red mudrocks, is abundant, averaging 282 ppm. Y concentrations discriminate between the oxic and anoxic sedimentary facies, being higher in red mudrocks and associated facies (average 46 ppm) than in black mudrocks and related coarser facies (average 20.3 ppm).

The clearest discrimination of the facies groups is shown by plots of $ZrATiO_2$ v. Nb/Y. Sandy turbidites, intercalated with the mafic sheets, virtually all fall in the andesite field. All the sand- and gravel-grade rocks, which are interbedded with red mudrocks, plot well within the rhyodacite-dacite field as is to be expected from the approximately equal quartz and feldspar phenocryst distribution in the clasts. Nb/Ti plots for the sand-and gravel-facies rocks show that the distribution of these elements even more clearly segregates samples in accordance with their petrography. The sandy turbidites and gravel-grade sedimentary rocks interbedded with the black mudrocks have Ti/ Nb ratios between 400 and 800 whereas the sand- and gravel-grade sedimentary rocks associated with the red mudrocks all have Ti/Nb ratios below 400. Additional data for the red mudrocks, and interbedded coarser clastic rocks, come from the trace-element lithogeochemical survey by Oswin (1986) of the Transition Series. His data for these sedimentary rocks show a very clear restriction in Ti/Nb ratios to less than 400:1.

Red and black mudrocks from the Volcano-Sedimentary Complex, along with a sample from the Phyllite-Quartzite Group, occupy the same general field on Ti/Nb plots, consistently having Ti/Nb ratios less than 400:1. Data from the Phyllite-Quartzite Group presented by Mullane (1998) show that material from this stratigraphic unit always falls in this field, being mainly restricted to Ti/Nb ratios between 200:1 and 400:1, and that Nb is generally abundant (most > 13 ppm). Similarly, the chondrite-normalised REE patterns of all the mudrocks essentially have the same shape with light REE (LREE) enrichment, flat heavy REE (HREE) segments, and significant negative Eu anomalies. (La/Lu)N ratios range from 6.49 to 10.31, except for sample

32, which has a ratio of 2.83. All of the REE analyses have Eu/Eu* ratios in a narrow range of 0.56-0.74. Some differences are also noteworthy.

Four samples have almost identical patterns. This group comprises sample 1 from the Phyllite-Quartzite Group several hundred metres below the Volcano-Sedimentary Complex; sample 22, a slate within the pebbly mudstones of Corta Atalaya ('Slate Conglomerate'); sample 28, from a stratified zone low in the thick felsic sheet SE of Rio Tinto; and sample 35, a hematitic slate c. 10 m above the massive sulphide layer in Corta Atalaya. Samples 20 and 23, intercalated with the andesitic turbidites in the Jarama River section, stand apart from the grouping of four noted above because of their lower total REE abundances. However, all six of these samples have comparable patterns, although samples 20 and 23 show lower LREE to HREE fractionation as illustrated by $(La/Lu)_N$ ratios of 6.65 and 6.84. The two remaining analyses are distinctive. Sample 29 has the lowest REE total and sample 32 has lower fractionation between LREE and HREE, which is reflected in a $(La/Lu)_N$ ratio of 2.83.

Plots of $ZrATiO_2$ v. Nb/Y and Ti v. Nb data for the primary volcanic rocks facilitate comparison with the associated sedimentary rocks. The igneous rocks have been subdivided into pre- and post-mineralisation groupings based on the field criteria established by Boulter (1996). The pre-mineralisation igneous rocks all come from the Rio Tinto stockwork, although they can be classified using elements relatively insensitive to alteration. It is evident that the spectrum of igneous compositions emplaced before the major exhalative episode at Rio Tinto ranged from rhyodacite to basaltic andesite. The post-mineralisation igneous rocks are distinctly bimodal with a conspicuous absence of intermediate compositions.

Ascertaining provenance of sedimentary rocks is commonly frustrated by a multiplicity of processes that operate on the materials between source and sedimentary rock, including chemical weathering, hydraulic sorting, diagenesis, metamorphism and hydrothermal alteration. Fortunately, for the coarse clastic rocks considered here, the predominance of rock fragments, having a limited range of compositions and textures, means that there will be a much closer relationship between sedimentary composition and the source than for materials that have undergone more extended sedimentary sorting, or chemical weathering before any transport. However, identifying the chemical signatures of the parent to the mudrocks is a much more challenging process because of the multiplicity of chemical-mineralogical modifications associated with the evolution from fresh source rock to lithified mud.

Chemical modification is intense in this process and only a limited range of elements, generally those with low to moderate seawater-upper-crust partition coefficients, are capable of reliably reflecting the source composition. For some elements within this selection the possibility of quantitative transfer from source to sedimentary rock is reduced by factors such as heavy mineral fractionation, changes of solubility with differing oxidation states, and chalcophile behaviour. For these reasons geochemical provenance studies typically concentrate on the Tigroup, the Al-group, the REE, Th, Nb, Sn and be as the most reliable means of deducing the nature of source terrains.

Bonjour & Dabard (1991) have shown that Ti/Nb ratios are particularly good provenance indicators in clastic sedimentary rocks because these elements undergo little fractionation during weathering or diagenesis. The utility of Ti and Nb in sedimentary provenance studies has also been demonstrated by Slack & Stevens (1994). Recent work has confirmed the lack of mobility of Ti and Nb during weathering but has shown, in some circumstances, significant mobilisation of Y and some perturbation Of Zr.

Major Elements

Based on the considerations discussed above, only generalisations can be made when using the bulk of the major-element chemistry. The lack of single linear arrays on two-component plots indicates a complex picture for the major elements. A linear pattern on element-element plots for sedimentary rocks would typically represent mixing lines generated by the sorting of clay-rich and quartz-rich components during sedimentation, given a reasonably homogeneous source. If this process does not involve gain or loss of material then the array will pass through the source composition. For the Rio Tinto area, the lack of a well-defined correlation between pairs of major elements could imply significant clement mobility during any one or some combination of the overprinting processes noted earlier. Alternatively, or additionally, several sources may be implicated. Despite the potential for chemical disturbance during alteration, a mixture of felsic to intermediate provenance characteristics for the major-element chemistry is evident, based particularly on MgO, and K_2O, concentrations.

Contrasts between Igneous and Overlying Sedimentary Rocks

Petrographic and geochemical contrasts between sedimentary rocks and intercalated, or stratigraphically lower, primary volcanic facies are common in the Volcano-Sedimentary Complex. Of these patterns, the alternation of intermediate and mafic signatures, through at least five sedimentary-rock-dolerite-sheet repetitions in the Jarama section, is one of the most prominent. Ti/Nb ratios, and to a lesser extent Zr/TiO_2 v. Nb/Y data, clearly discriminate between the mafic sheets and the intercalated stratified rocks. Sedimentary packages with average Ti/Nb ratios of 512:1 alternate with mafic sheets that have average Ti/Nb ratios of 1142:1. It is difficult to explain these pronounced contrasts in terms of volcanic-pile processes. If the mafic sheets were effusive and associated with primary volcaniclastic rocks, they should have contributed detritus to any immediately overlying sediments.

Even in volcanic settings with subdued topography, such as mafic submarine lava plains, examples have been documented where intercalated sedimentary horizons have been entirely sourced in the mafic rocks. In comparable settings, the provenance of intercalated sedimentary rocks is not exclusively the local volcanic rock but its geochemical signature is clearly identifiable. The likelihood of local contributions to sedimentary provenance is increased by the recognition that deep-water volcaniclastic deposits are much more extensive than previously recognised and that deep-marine current activity is a significant contributor to the nature of these deposits. For the Jarama River section, the mafic sheets did not supply detritus to the overlying sedimentary rocks, as is demonstrated by the petrographic and chemical contrasts. The volcanic-pile model is not supported by these data, which instead favour intrusive behaviour for the mafic sheets as implied by the abundant field observations of widespread interaction between magma and wet sediment. Stratified rocks of a mafic composition are not recorded in the region and it would appear that the peperitic mafic intrusions did not break their cover to contribute to the sediment source.

Additional information on the part of the Jarama section that contains the mafic sheets comes from the REE data on the mudrocks (samples 20 and 23) associated with the andesitic turbidites. These data, in pattern and absolute abundances, closely match REE data from Iberian Pyrite Belt andesites. Basalts and basaltic andesites in the Iberian Pyrite Belt have flat REE patterns and do not have a negative Eu anomaly. This is very different from the patterns in the mudrocks intercalated with the mafic sheets in the Jarama section. Therefore, the mudrocks between the mafic igneous sheets have andesitic characteristics and these samples further highlight the lack of mafic rock provenance, thus supporting the field textural evidence for an intrusive mode of emplacement for the mafic sheets.

A more pronounced contrast, between the primary-volcanic-facies substrates and overlying stratified rocks, is represented by the alternation of mafic sheets and felsic volcaniclastic rocks in the Odiel River. Here the primary mafic volcanic facies alternate with felsic conglomeratic facies. The cobble-grade material in these sedimentary rocks is quartz- and feldspar-phyric and the bed characteristics arc comparable with the debrite zone towards the top of the Jarama section. The mafic sheets have not contributed at all to the felsic volcanic-derived volcaniclastic breccias, thus supporting an intrusive origin for the mafic volcanic rocks.

Geochemical data for the felsic igneous sheets within the post-mineralisation bimodal suite and the coarse clastic rocks in the red mudrock part of the stratified sequence show considerable overlap in their range of compositions. It is therefore not possible to use the geochemistry obtained in this study to link particular felsic sheets to their derived hydrovolcanic eruption products. Textural characteristics are more distinctive than chemical variations, but mingled magmas, and the composite nature of some sheets, limit the reliability of correlations in sections where both primary and stratified facies are felsic.

Persistence of Continental Signatures Through the olcano-Sedimentary Complex

REE data from mudrocks contribute to the identification of continental-derived components within the Volcano-Sedimentary Complex. The cluster of four nearly identical mudrock REE patterns is indistinguishable from the post-Archaean shale composites presented by Taylor & McLennan (1985) and considered by them to represent the upper continental crust for the REE although with a slight overestimate of absolute abundances. The (La/Lu)N ratios for this group range from 9.5 to 10.31 and compare well with the value of 9.2 for this ratio for the Post-Archaean Australian Shale of Taylor & McLennan (1985). All of these mudrocks would appear to be representative samples of average upper continental crust. Such a conclusion was to be expected for the Phyllite-Quartzite Group sample 1 because this is stratigraphically well below the Volcano-Sedimentary Complex.

In contrast to sample 1, samples 22, 28 and 35 all are located well within the Volcano-Sedimentary Complex, and their upper-crustal affinity poses considerable difficulties for the volcanic-pile model because volcanic substrates should contribute significantly to the overlying clastic rocks. The mudrock sample from the 'Slate Conglomerate' in Corta Atalaya (sample 22) is one of the closest matches to the Post-Archaean Australian Shale, yet, in the volcanic-pile model, it had formed after one-third to one-half of the Volcano-Sedimentary Complex. The REE data coupled with the pebble compositions create the most pronounced contrast between primary volcanic facies and overlying stratified rocks in the region. Schermerhorn invoked exotic explanations by describing them as 'a flow tilloid deposited from a mudflow descending into and coming to rest in a deeper part of the basin, among volcanics'. In the sill-sediment-complex scenario, the sequence of pebbly mudstones is viewed as forming before and partially overlapping the first peperitic intrusions in the region such that volcanic contributions were minimal, and therefore intercalated mudrocks would not have detectable volcanic chemical signatures. Volcanic textures in the andesitic to felsic rocks have largely been obscured by the stockwork-related alteration but preservation of some peperitic textures supports an intrusive origin.

Although sample 28 would be considered in the volcanic-pile stratigraphy to be above the Basic Complex and to be in the lower part of the Acid Complex, its composition closely matches the post-Archaean shale composites. It is inferred that the increase in thickness has meant that the sill has cut to deeper levels than elsewhere into the pro-intrusion Phyllite-Quartzite Group and included rafts of that unit. This accounts for the Post-Archaean Australian Shale-type composition of sample 28. Other characteristics of the adjacent stratified rocks, such as intraformational conglomerates and pre-lithification deformation, match the features of the uppermost Phyllite-Quartzite Group.

The upper-crustal REE pattern from the red bed sequence in Corta Atalaya (sample 35) is highly anomalous in the volcanic-pile model because it is part of the Transition Series that post-dates the bulk of the Volcano-Sedimentary Complex (the Basic and Acid Complexes) and would therefore be expected to show a volcanic signature. Sample 35 is intercalated with sandy turbidites on the several millimetres to few centimetres scale that comprise less than 10% of this part of the succession. As is the case for the other oxic-facies volcaniclastic rocks, the arenites are felsic. Volcaniclastic input into the sequence above the sulphide sheet appears to have been much less than in lateral facies adjacent to the mineralised area and this is mirrored in the mudrock chemistry. Because the primary volcanic facies that predate the mineralisation are restricted to the immediate vicinity of the sulphide deposits, it is possible that a positive topography had been created in these areas by the time exhalative activity had ceased. Deformation of the sea floor above the high-level intrusions, spatially related to the extensional faults, may have reversed the tectonic depression of the basin floor during the formation of the half-graben. The large debris flows associated with the red mudrocks appear to have been restricted to the lower-lying areas that had no, or minimal, primary volcanic facies until after mineralisation. Comparable hematitic mudrocks that cap the massive sulphide bodies at Aljustrel also have no volcanic aspects to their geochemistry.

Sample 32 is a red mudrock from the Jarama section that is laterally equivalent to the sample 35 locality. This sample is from the part of the section where debris-flow deposits dominate and the volcanogenic input can also be seen in the mudrock REE pattern, which is comparable with data published by Mitjavila et al. (1997) for dacites in the Iberian Pyrite Belt. Where the red mudrock succession is dominated by major volcaniclastic input as evidenced by beds up to 30 m thick, the mud may have been largely generated during the hydrovolcanic process, possibly by spalling of glassy material as fluidized sediment came into contact with cooling igneous material along fracture planes. In contrast, the mineralised zones, above the topography created by the pre-mineralisation igneous rocks, were not affected by this catastrophic sedimentation. These mud-dominated areas received only a little volcanogenic detritus in the form of very thin turbidites, and the background pelagic sedimentation was of a regional, continental, provenance as shown by the Post-Archaean Australian Shale-type REE data.

The Volcano-Sedimentary Complex contains many examples of pronounced compositional contrasts between stratified rocks and underlying primary volcanic facies. Particularly striking are the contrasting compositions between alternating, largely coherent, mafic sheets and volcaniclastic turbidites-debrites. In the Jarama River the mafic sheets alternate with stratified rocks of andesitic composition where distinct populations of Ti/Nb ratios highlight the chemical contrasts. In the Odicl River, the mafic sheets alternate with bedded volcaniclastic rocks of a felsic composition. Perhaps a more significant contrast is found in Corta Atalaya between the virtually non-volcanic pebbly mudstones stratigraphically overlying hundreds of metres of intermediate, and felsic, primary volcanic rocks. These relationships cannot be reconciled with the provenance patterns expected for a continuously evolving volcanic pile where the underlying primary volcanic facies should be available to contribute to any sedimentary horizon formed in a volcanically quiescent period. REE geochemical signatures of a continental style are distributed throughout the Volcano-Sedimentary Complex stratigraphic column, supporting the proposal that much of the magmatic activity was confined, i.e. not eruptive. That this was not exclusively the case is demonstrated by the volcaniclastic turbidites and debrites of intermediate and felsic compositions. The andesitic component of the stratified rocks was derived from localised areas of magmatism centred on zones that became sites of exhalative hydrothermal activity. In contrast, the mafic part of the post-mineralisation bimodal intrusive suite does not appear to have contributed to any of the stratified volcaniclastic rocks in the Rio Tinto district, and appears to have been totally confined. The test provided by the sedimentary provenance supports the field evidence for a major role for peperitic intrusions in the facies architecture of the host to one of the world's most significant mctallogenic provinces.

Confirmation of the sill-sediment-complex nature of the Volcano-Sedimentary Complex highlights the need to reassess the physical volcanology of this volcanic unit to provide a basis for the ore-deposit modelling. Reconstructions of the synmineralisation setting will be in error if they arc based on an inappropriate model for the physical volcanology. Other aspects of the metallogeny affected include the modelling of the hydrothermal system, the understanding of the alteration patterns, the determination of metal sources, mechanisms of ore-body segmentation, correlating the ore-equivalent horizon, and discriminating pre- and post-mineralisation igneous rocks. All of these topics also have relevance for exploration methods. It is vital that the physical volcanology of this region is well defined, but it seems certain from data presented here that the volcanic-pile model is invalidated and there is strong support for the sillsediment-complex interpretation.

Geochemistry of Coeval Mesozoic Plutonic

High precision U-Pb zircon age-dating of volcanic and plutonic rocks in Hong Kong (herein the area of the Hong Kong Special Administrative Region) has placed tight constraints on the timing of Jurassic to Cretaceous magmatic events in a small area (1050 km^2) of the SE China igneous province. The data reveal four discrete episodes of magmatism, generally of less than one million years duration, separated by longer periods of volcanic and plutonic quiescence. Such precise dating, together with detailed field, petrographic, geochemical and isotopic data, provides an opportunity to examine the close interrelationships between volcanism, plutonism, and tectonism. This chapter describes the compositional variation through time of Hong Kong volcanic and plutonic suites and discusses aspects of their petrogenesis and tectonic setting.

Geological Setting

Destructive plate magmatism along the Eurasian continental margin during the Mesozoic Yanshanian Orogeny culminated in voluminous intermediate to silicic eruptions and the emplacement of numerous, high-level granitoid plutons. These Jurassic to Cretaceous volcanic and plutonic rocks crop out over a broad (400 km) belt along the coastal region of SE China and record a complex tectonomagmatic history.

The coastal region of SE China is dominated by several large-scale NE-trending fault zones the most prominent of which is the Lianhuashan Fault Zone extending for over 200 km through the maritime provinces of eastern Guangdong, Fujian and Zhejiang. In eastern Guangdong, the Lianhuashan Fault Zone comprises a 30 km wide zone bounded to the north by the Shenzhen Fault and to the south by the Haifeng Fault. Numerous volcanic centres and plutonic assemblages within this zone, including those of Hong Kong, suggest that the Lianhuashan Fault Zone was a major locus of magmatic activity throughout the Late Mesozoic. This is corroborated by Sr and Nd isotope studies of the granitoid rocks, geophysical studies and zircon inheritance studies, which collectively endorse the existence of a deep crustal boundary through Hong Kong separating dominantly late Archaean crust in the NW from Proterozoic crust in the SE.

Approximately 85% of the area of Hong Kong is formed from Mesozoic volcanic-plutonic assemblages. The bulk of these rocks are of Mid-Jurassic to Early Cretaceous age and are divided into two volcanic groups and two granitoid suites. During the course of this study it has been necessary to introduce a refinement of this suite and group nomenclature in order to discuss the close compositional and temporal relationships among the lithological units. Accordingly, the granitoids are grouped in terms of magmatic pulses where a 'Pulse' is defined as a magmatic event in which one or more single intrusive units (plutons) with similar bulk chemistry and mineralogy were emplaced at the same time. Similarly, the volcanic units are grouped in terms of phases where a 'Phase' is the volcanic equivalent of a 'Pulse'. A period (or episode) of magmatic activity may comprise one or more pulses or phases.

Volcanic and Plutonic Associations

The Middle Jurassic-Lower Cretaceous volcanic and plutonic rocks of Hong Kong are both spatially and temporally related. Subvolcanic dykes, stocks, and plutons also commonly have whole-rock geochemical signatures similar to those of the volcanic formations making it possible to link compositionally the plutonic pulses to one or other of the volcanic phases. The broad compositional similarities between volcanic rocks and granitoids from the four main episodes of magmatism are demonstrated on a plot of TiO_2 versus Zr. However, within each of these episodes, the relationship between volcanic phases and plutonic pulses is by no means straight forward and not all intrusions have subvolcanic equivalents and vice-versa.

Volcanic rocks of the Tsuen Wan Volcanic Group and granitoids of the Lamma Suite are mostly exposed in the north and west of Hong Kong. Based on U-Pb zircon age-dating and whole-rock geochemistry, the Lamma Suite granitoids can be divided into two groups, here termed Pulse 1A and Pulse 1B. Pulse lA granitoids were emplaced between 164.6 and 161.5 Ma and are temporally and geochemically related to the volcanic rocks of the Tsuen Wan Volcanic Group (Phase 1). Granitoids of Pulse lB have yielded ages between 160.4 and 159.3 Ma but do not have a recognisable volcanic equivalent. In terms of whole-rock geochemistry, Pulse IA granitoids have distinctive I-type characteristics whereas those of Pulse IB have A-type characteristics. Based on the new U Pb age data, two granitoids previously assigned to subgroup 1 of the Lion Rock Suite are now included with Pulse IB of the Lamma Suite.

A significant break (13 Ma) exists between the first and second episodes of Mid-Jurassic-Early Cretaceous magmatism. Volcanic and granitoid rocks belonging to this second episode were emplaced between 146.6 and 146.2 Ma. They are designated as Phase 2 and Pulse 2 respectively, and are largely restricted to a 6 km wide, NE-trending zone through central parts of Hong Kong. These rocks comprise an assemblage of dominantly I-type extrusive and intrusive units, but include a group of late stage A-type plutons and dykes for which there are no known volcanic equivalents.

Two compositionally distinct, intercalated, volcanic plutonic assemblages have been identified within the third episode of magmatic activity which took place between 143.7 and 142.7 Ma. These comprise a large volume of dacite to rhyolite volcanic rocks and I-type monzogranites and quartz monzonites (Phase 3/Pulse 3), and a smaller volume of trachyte to rhyolite volcanic rocks and associated quartz monzonite intrusions (Phase 4/Pulse 4). Both assemblages are exposed mainly in the south and east of Hong Kong and the granitoids have distinctive I-type compositions.

Volcanic and granitoid rocks belonging to the fourth main episode of magmatism (Phase 5/Pulse 5) were emplaced between 140.9 and 140.4 Ma. They are largely concentrated in the southeast of Hong Kong and comprise thick, caldera-related silicic tuff sequences, two large monzogranite plutons and several smaller quartz monzonite stocks and dykes. The volcanic rocks are dominantly rhyolite in composition and the granitoids vary from fractionated I-type compositions to A-type compositions.

The final pulse of Mid-Jurassic-Early Cretaceous silicic magmatism is recorded by a number of small quartz monzonite to monzogranite stocks and dykes in the south of Hong Kong. These rocks intrude monzogranite of Pulse 5 but have not yet been radiometrically dated. They are volumetrically minor by comparison with the main plutonic assemblages and may in fact belong to a late intrusive stage of the fourth episode. These rocks are tentatively assigned to Pulse 6 and have no known volcanic equivalents.

General features of the mineralogy and petrography of individual formations and plutons are described elsewhere and only a brief summary is necessary here. The Tsuen Wan Volcanic Group (Phase 1) is composed mainly of coarse ash crystal tuff although significant volumes of tuff-breccia, lapilli tuff. volcaniclastic sedimentary rocks and fine ash crystal tuff are also present. Coarse ash crystal tuff is the dominant lithology of the Repulse Bay Volcanic Group (Phases 2-5) but significant volumes of fine ash vitric tuff, which is commonly welded, is also present and form thick ignimbritic sequences. Intercalated volcaniclastic sedimentary rocks have been recognised in many units of the Repulse Bay Volcanic Group whereas lava flows are relatively uncommon. The mineralogy of the volcanic rocks closely resembles that of the related plutonic rocks.

The I-type Lamma Suite (Pulse IA) granitoids are commonly K-feldspar megacrystic and medium grained, and contain strongly zoned plagioclase, with quartz, glomerocrystic biotite, and minor amphibole. Accessory minerals include magnetite, zircon, zoned allanite, and apatite. A-type granitoids (Pulse 1B) comprise fine grained biotite monzogranite and leucogranite the latter containing fluorite, sparse muscovite, and rare garnet. Rhyodacite and rhyolite porphyry dykes of Pulse 2 (Lion Rock Suite) contain large (5 20 mm) euhedral K-feldspar megacrysts along with subordinate bipyramidal quartz. zoned plagioclase, and biotite glomerocrysts set in a fine-grained granophyric matrix. Accessory minerals are similar to those of the Pulse I I-type granitoids. Equigranular biotite monzogranite is the dominant lithology of Pulse 3 and Pulse 5 granitoids. Plagioclase in these rocks is strongly zoned, and accessory minerals include magnetite, zircon, titanite, allanite and fluorite. Quartz monzonites of Pulse 3 and Pulse 4 are dominantly porphyritic and medium grained. The main mafic minerals are hornblende and biotite, and accessory minerals include zircon, apatite, and magnetite. Quartz monzonites and monzogranites of Pulse 6 are principally K-feldspar porphyritic and fine grained. Biotite and amphibole are the dominant mafic constituents while the accessory minerals include magnetite, zircon, titanite and fluorite.

Major and trace element composition on a plot of K_2O v. SiO_2, almost all of the analysed rocks fall within, or above, the high-K calc-alkaline field of Le Maitre (1989). Many samples are extremely K-rich and plot in the field for fractionated alkaline shoshonitic series compositions.

Phase I rocks show slightly increasing K20 abundances with increasing degree of differentiation whereas the rocks of Phases 2 to 5 show flat, or slightly decreasing abundances with increasing SiO_2 content. Some of the anomalously high K_2O contents shown by Phase 2 volcanic rocks correspond to those which fall outside the normal igneous spectrum. These K_2O enriched rocks may have been affected by low temperature hydration and devitrification of rhyolite glass.

Granitoids of Pulses 1A and 1B show similar K_2O abundances and fractionation trends to volcanic rocks of the Phase I with compositions plotting within the zone bounding shoshonitic and high-K calc-alkaline series compositions. K_2O contents and fractionation trends for Pulse 2 rocks are also similar to those of the least-altered Phase 2 rocks, and the same can be shown for granitoid and volcanic rocks of Pulse 3/Phase 3. There are small differences in the K_2O contents and fractionation trends between Pulse 4 granitoids and Phase 4 volcanic rocks, with the volcanic rocks showing a trend of slightly increasing KO with degree of differentiation compared to the slight decreasing trend shown by the granitoids. The variation in K_2O abundances in Pulse 5 granitoids is similar to that of Phase 5 volcanic rocks. Pulse 6 granitoids show slightly higher K_2O contents than Pulse 5. Overall, K_2O contents for rocks of Phases 2 to 5 and Pulses 2 to 6 are higher than for the Phase 1 and Pulse 1 rocks and show a progressive change toward shoshonitic compositions with time.

Trace Element Chemistry

General fractionation trends are displayed on 'Harker' variation diagrams of selected compatible and incompatible elements. Depletion of Sr with increasing degree of differentiation is consistent with the fractionation of plagioclase and apatite as indicated by the major element plots. The broader scatter of data points in some of the volcanic phases reflects the mobility of Sr during alteration processes. These samples also generally show the most variable K_2O and Rb contents. From 63 to 75 wt%, SiO_2, systematic increases in elemental abundances shown by Rb and Nb indicate that these elements behaved incompatibly. Extreme enrichment in these elements is shown by the most fractionated (> 76 wt"/SiO_2) granitoid samples.

A plot of Ti versus Zr is particularly useful for discriminating the various volcanic phases and plutonic pulses. Separate evolutionary paths are clearly defined with the least-evolved members of each group having distinct initial Zr/Ti ratios. Within each of the paired phases and pulses, both elements behave in a similar manner during fractional crystallisation. The steep trends in groups 1, 2, and 3 indicate that Ti is behaving more compatibly than Zr, whereas the shallower trends indicate that both elements have similar compatible behaviour. Differences in the rate of removal of these elements during fractionation are likely to be controlled by crystallisation of zircon and Ti magnetite. The available Nd and Sr isotope data rule out significant high-level crustal contamination. Differences in the initial Zr/Ti ratios for each group, therefore, are thought mainly to reflect differences in the composition of the source material.

The separate fractionation lineages displayed by each of the volcanic phases indicates a step-like evolutionary change towards Zr-enriched, Ti-depleted compositions with time. For example, Phase 3 compositions are less Ti-rich than Phase I but have higher Ti contents than Phase 2 at similar degrees of differentiation. Furthermore, there is marked compositional distinction between volcanic rocks of Phases 3 and 4 which were erupted during the same 143 Ma magmatic episode. This strongly suggests that magmas were being derived from two sources coevally. A similar, though less obvious distinction in source composition among coevally erupted magmas is shown for Phase 5 where two subphases can be distinguished: an early trachydacite-rhyolite subphase (SA), which is intermediate between Phases 3 and 4, and a late trachyte rhyolite subphase (5B), which is slightly less Ti-rich than Phase 4. The broad scatter of data points in Phases 1, 2, and 3 is partly due to subordinate evolutionary trends represented by individual formations.

The granitoids preserve patterns of Zr-Ti variation similar to those of the volcanic rocks although among the strongly differentiated compositions the fractionation lineages are less apparent. In general, Pulse 1 granodiorites and monzogranites, have very similar trends and abundances to volcanic rocks of Phase 1, although there is no volcanic equivalent for the Pulse lB monzogranites. Monzogranites and dykes of Pulse 2 also have similar Zr-Ti contents and fractionation trends to the volcanic rocks of Phase 2. Among the quartz monzonites, those of Pulse 3 are the most Ti-rich and may be more primitive members of the Pulse 3 monzogranite lineage. Comparing Pulse 3 granitoids with Phase 3 volcanic rocks, there is broad correspondence in both Zr Ti abundance and fractionation trend. Quartz monzonites of Pulse 4 have very similar Zr-Ti contents to Phase 4 volcanic rocks but lower Ti abundances compared to the coeval quartz monzonites of Pulse 3. Quartz monzonites of Pulse 5 can be split into an earlier pulse (SA) and a later pulse (5B) which broadly correspond with the volcanic subphases (SA and 5B). Whereas quartz monzonites of Pulse SA lie on a fractionation trend that would give rise to the monzogranites, those of Pulse 5B are distinctly less Ti-rich, and lie on a fractionation lineage corresponding to a small group of slightly younger, less Ti-rich monzogranites. Although no volcanic equivalents have yet been identified for Pulse 6, these rocks record the lowest Ti and highest Zr contents of any magmatic group in Hong Kong. There is a possible fractionation lineage linking the quartz monzonites and monzogranites of this youngest recognised pulse. On MORB-normalised spider diagrams, representative compositions from the five phases of volcanic rocks have consistent profiles that are similar to those magmas found in active continental margins. In particular, the profiles show enrichment of K, Rb, and Th, and depletion of Nb, which are characteristic features of subduction-related magmas and an indication of crustal involvement in magma genesis. Troughs for Ba and Sr can be explained by fractional crystallisation of feldspar whereas the negative Ti anomaly shown by all the analysed compositions reflects the crystallisation of Ti-magnetite.

MORB-normalised trace element patterns for representative granitoids from each pulse are similar to those of the volcanic rocks and a similar relationship is exhibited by Phase 2 and Pulse 2 rocks on a chondrite-normalised plot of the REE data. The chondrite-normalised plot also highlights the similarities in REE behaviour among the Middle Jurassic Lower Cretaceous granitoids. In general, from Pulse I through to Pulse 5, REE abundances increase while preserving the light REE-enriched shape of pattern. Sewell et ail (1992) have drawn attention to the rather flat to slightly heavy REE enriched patterns shown by the A-type granitoids of Hong Kong which have overall REE concentrations of up to sixty times chondrite abundance. The quartz monzonites commonly show a slight negative dip in the middle REE pattern which is thought to be caused by amphibole fractionation. The Middle Triassic Deep Bay Granite has markedly lower heavy REE contents compared to the Jurassic-Cretaceous granitoids ruling out a genetic relationship.

The general pattern of evolution among the Mid-Jurassic Early Cretaceous magmas is for the least-evolved compositions of each group to become more K_2O enriched, Zr-enriched and Ti-depleted with time. Separate batches of magma generated at 164 159 Ma, 146 Ma, 143 Ma, and 140 Ma were derived from variable degrees of partial melting of crustal material involving a mantle-derived component. Each batch of magma was characterised by a slightly different initial Zr/Ti ratio and differentiated mainly by fractional crystallisation involving the removal of plagioclase, K-feldspar, biotite, amphibole and Ti-magnetite.

Darbyshire & Sewell have isotopically modelled the sources of Hong Kong granitoid magmas in terms of a mantle component and two crustal components. Based on Sr and Nd isotope data, they propose that the source of the Lamma Suite granitoids (Pulse 1) in the northwest of Hong Kong is a mixture between a juvenile mantle source and dominantly Archaean crustal material whereas in the southeast of Hong Kong, the source of the Lion Rock Suite granitoids is a mixture between a mantle-derived melt and dominantly Proterozoic crustal material. They interpret the variable isotope signatures in the area between as marking the boundary between these two dominant crustal sources and recording the strongest mantle-derived imprint.

The overall consistency of REE patterns and abundances among the I-type granitoids of Hong Kong is in marked contrast to the differences in the Nd isotope initial ratios. One explanation for this disparity could be that the isotope signatures are more sensitive indicators of variations in the crustal source component. This situation could arise if the isotope signatures of the crustal component did not have sufficient time to re-equilibrate following assimilation by juvenile mantle-derived melts. Alternatively, the REE patterns and abundances of the I-type granitoids may largely reflect those of the mantle-derived component. The strongly peraluminous composition and negative ENd signature (-13.9) of the Middle Triassic Deep Bay Granite preclude an appreciable mantle-derived component in the source region. In this instance, the distinctive REE pattern of the Deep Bay Granite is considered to reflect that of the dominant Archaean crustal source component. If the REE contents of the I-type granitoids were also primarily related to those of the crustal component, then their REE patterns would be expected to vary in concert with their isotope signatures. Since this is not the case, the consistent REE patterns and abundances displayed by the Upper Mesozoic rocks are thought to largely reflect those of the mantle-derived component.

The involvement of both Archaean and Proterozoic crust in the genesis of Hong Kong magmas has largely been confirmed by the presence of inherited zircons in the granitoids and related volcanic rocks. Together with the REE evidence for a consistent mantle-derived signature, we propose that the variability in magmatic composition is best explained by interaction of a mantle-derived melt with differing proportions of Archaean and/or Proterozoic crustal material.

The higher $^{87}Sr/^{86}Sr$ initial ratios and lower õ Nd signatures of Pulse I granitoids compared with the other groups suggest that mantle-derived melts interacted strongly with an Archaean-dominated crustal source. A strong crustal and mantle influence is also preserved in the source area of Pulses 4 and 5 although the isotope data indicate that the crustal contaminant is likely to have been dominantly Proterozoic in age. The strongest mantle influence is preserved in the sources of Pulse 2 and Pulse 3 granitoids. Geographically, both groups lie within the Zone 2 of Darbyshire & Sewell which has been interpreted as representing a region of deep crustal faulting. These discontinuities of crustal dimensions may have promoted the mixing of mantle-derived magmas with both Archaean and Proterozoic crustal sources.

Campbell & Sewell have shown that there are fundamental structural controls on the distribution and disposition of volcanic centres in Hong Kong which are typical of volcanism within the Lianhuashan Fault Zone. They conclude that the onset of transtension and extension in the Mid- to Late Jurassic and Early Cretaceous led to the development of large calderas and high-level subvolcanic magma reservoirs. Lithologically, the volcanic rocks of Phase 3 are mainly composed of coarse ash crystal tuff of dacite to rhyolite composition. The crystal-rich character of these deposits suggests the development of high-level, subvolcanic magma chambers which permitted prolonged crystal fractionation prior to eruption. These magma chambers are now preserved as large elliptical plutons (Pulse 3) which have demonstrable links to the overlying volcanic rocks. By contrast, most of the volcanic rocks belonging to Phase 4 are trachyte to rhyolite in composition and their plutonic equivalents occur mostly as quartz monzonite dykes. The volcanic deposits of Phase 4 are dominantly composed of fine ash vitric tuff, commonly thick and densely welded indicating that they were emplaced rapidly and at high temperature. These deposits mainly accumulated adjacent to linear fissure vents. Thus, two main types of volcanic plutonic assemblage can be recognised: 'caldera-type' in which volcanic deposits accumulated in tectonomagmatic depressions controlled mainly by ellipsoidal and rectilinear fracture systems, and 'fissure-type' in which volcanic deposits accumulated adjacent to fissure vents controlled by linear fracture systems. The two best examples of 'fissure-type' plutonic-volcanic assemblages include the Phase SA assemblage and the Phase SB assemblage, whereas the two best examples of caldera-type' assemblages include the Phase 3 assemblage and the Phase lA assemblage. It is, therefore, conceivable that the low Zr/Ti, trachyte-rhyolite 'fissure-type' magmas migrated relatively rapidly to the surface along deep crustal pathways and did not develop large subvolcanic reservoirs.

Caldera development, coupled with a transcurrent movement, is a feature of some back-arc continental settings. Campbell & Sewell postulate that the emplacement style and character of the Phase I volcanic rocks and related Pulse 1 intrusions suggest the incipient development of a back-arc setting in which extension and transtension are the dominant tectonic controls. Such a setting is likely to promote the development of high-level, magma storage reservoirs and associated calderarelated volcanic activity. Extension and transtension can also be regarded as the main tectonic controls on the development of 'caldera-type' volcanic-plutonic assemblages in Hong Kong, especially where relatively rapid extension can be demonstrated. Such rapid extension may have promoted the passage of large volumes of strongly mantle-influenced magmas to the surface. On the other hand, transtension may have promoted the interaction of mantle-derived magmas with continental crust resulting in the generation of increasingly K_2O-rich and TiO,-depleted magmas. These magmas then rose relatively rapidly to the surface through transtensional crustal pathways to be extruded predominantly along linear crustal fissures.

Precise dating of the Mesozoic volcanic and plutonic rocks of Hong Kong has placed very tight controls on the sequence of magmatic events. Combined with stratigraphic, isotope and geochemical data, it has been possible to demonstrate the existence of co-magmatic relationships among coeval volcanic and plutonic pairs indicating an inextricable link between eruptive products and their subvolcanic sources. The earliest calc-alkaline andesitic eruptives are likely to represent a remnant of an extensive andesitic volcanic arc province that dominated the Eurasian plate margin during the Early to Mid-Jurassic. Although these arc-type rocks have not been precisely dated, the marked change in composition between these and the Middle Jurassic Phase 1 rocks suggests that either the intervening rocks have been eroded away, or that there was a period of relative volcanic quiescence in this part of the southeast China igneous province. Renewed volcanic activity of compositionally different character suggests that the westward-dipping subduction zone may have migrated farther east producing high-K calc-alkaline compositions distinctive of a mature continental arc province. This period of Mid-Jurassic to Early Cretaceous magmatism was punctuated by short bursts of activity during which the overall tectonic setting changed from mature continental arc to back-arc. Periods of rapid extension-dominated tectonism promoted the development of calderas, dyke injection, and high-level subvolcanic magma reservoirs. Periods of transtension promoted the magmatic activity along linear fissures.

The increasing K20-rich character of the magmatic products possibly reflects further eastward migration of the subduction zone as the depth to the Benioff zone increased. Shoshonitic compositions represent the final products of the Mid-Jurassic-Early Cretaceous magmatism prior to sudden termination at 140 Ma.

Geochemistry of Pelagic and Hemipelagic Carbonates

Recent work on Upper Cretaceous sequence stratigraphy in European basins has demonstrated that chalks provide valuable information on short- and long-term sea-level change. Sedimentary relationships on basin margins, combined with stratigraphic variation in lithology and carbon isotopes, have been used to define sequence boundaries, onlap surfaces and systems tracts. However, work on other Mesozoic sequences suggests that the elemental composition of pelagic sediments also varies in response to sea-level change.

Perhaps the best documented eustatic rise in sea level in Earth history occurred during the Cenomanian Stage with the global expansion of epicontinental seas and the widespread establishment of pelagic and hemipelagic carbonate deposition. The Cenomanian stratotype was used by Haq et aL (1988) to illustrate sequence stratigraphic analysis of an outcrop section, and a detailed sequence stratigraphic framework has been proposed recently for the Cenomanian of the Anglo-Paris Basin by Robaszynski et aL (1998). In this chapter we present the first systematic study of the elemental geochemistry of a Cenomanian Chalk succession. We test relationships between elemental and carbon isotope chemostratigraphy and the previously established sequence stratigraphy, and demonstrate that geochemical criteria provide a valuable tool for the recognition of sea-level change.

Culver Cliff on the eastern Isle of Wight, southern England, provides complete exposure of Cenomanian to Campanian chalks, traditionally assigned to the Lower, Middle and Upper Chalk formations. The section offers fresh exposures of pale green (unoxidised) sediments which have suffered only modest burial diagenesis, making them particularly well-suited to geochemical studies. The Cenomanian-Turonian (Lower and basal Middle Chalk) section was described by Kennedy (1969) and parts were studied more recently by Gale and Paul et al. (1994), who correlated these with various localities in NW Europe. The entire section was logged by us in detail and key litho- and biostratigraphic marker beds identified. As with other northern

European sections, a hiatus representing the uppermost Albian-lowest Cenomanian occurs at the base of the Chalk (Upper Greenland-Glauconitic Marl boundary), but the Cenomanian-Turonian succession at Culver compares favourably with those exposed elsewhere in the region.

Sequence Stratigraphic Framework

The Cenomanian of NW Europe typically consists of chalks and marls, rhythmically bedded on a decimetre-scale. This rhythmicity has been attributed to Milankovitch climatic forcing, probably driven by the 20 ka precession cycle, modulated by the short eccentricity cycle (100 ka bundles). However, lower frequency, tens of metres thick cycles are also recognised. Key horizons in these cycles have been correlated to onlap surfaces on the basin margin and locally on Palaeozoic massifs, providing a basis for placing these low-frequency cycles in a sequence stratigraphic framework. The Cenomanian is particularly well suited to sequence stratigraphic analysis because biostratigraphic resolution based principally on ammonites, is very high, enabling precise correlation of strata on a basinal scale.

Based on stratal geometry and variation in elastic input, Robaszynski et al. (1998) proposed six depositional sequences for the Cenomanian, each in the order of 1 Ma duration. Systems tracts were characterised and defined by diverse criteria including the geometry of sedimentary bodies, basin margin onlap, discontinuity surfaces (omission surfaces and hardgrounds) and lithological characters, particularly variations in quartz sand content, clay: carbonate ratios, and the presence of authigenic minerals, glauconite and francolite.

Eight major discontinuity surfaces that have been traced throughout the Anglo-Paris Basin, combined with other key marker horizons recognised in East Kent and the Boulonnais, may be used to correlate Cenomanian sections throughout southern England. These discontinuity surfaces and marker beds were used to place the sequence boundaries (SB), transgressive surfaces (TS) and maximum flooding surfaces (MFS) defined by Robaszynski et al. (1998) in the Culver Cliff succession. These surfaces define the positions of lowstand, transgressive, and highstand systems tracts which were assigned sequence numbers according to the scheme of Robaszynski et al. (1998).

Sequence I The basal Glauconitic Mart contact with the Gault Clay in East Kent and the Upper Greensand on the Isle of Wight, represents low (but not lowest) Cenomanian Neostlingoceras carcitanense Subzone, Mantelliceras mantelli Zone, resting disconformably on high Albian Stoliczkaia dispar Zone. The contact (DSI) is coincident with the sequence boundary/transgressive surface at the base of Sequence 1.

The maximum flooding surface of Sequence 1 is located in the basal Chalk Marl below a series of beds containing abundant Inoceramus crippsii Mantell bivalves. These beds occur < 1 m above the top of the Glauconitic Marl at Culver.

Sequence 2 Nodular limestone M3 of Gale (1989a) contains common Schloenbachia varians (J. Sowerby) and other ammonites typical of the uppermost N. carcitanense Subzone. The top of the bed is a prominent omission surface (DSII) overlain by a dark mart piped downwards by large Thalassinoides burrows. The omission surface marks the boundary between sequences 1 and 2, and represents a sequence boundary overlain directly by a transgressive surface.

The maximum flooding surface of Sequence 2 is not clearly defined in East Kent. At Culver we place it in the lower Chalk Marl, within a series of beds containing abundant Exanthesis labrosus (T. Smith) sponges.

Sequence 3 Phosphatic dark silty marl M4 lies within the Mantelliceras saxbii Subzone of the M. mantelli Zone, I-2 rn below the base of the Mantelliceras dixoni Zone. The discontinuity at the base of this marl (DSIII) marks the transgressive surface of Sequence 3, with the sequence boundary occurring below, in the immediately underlying many interval. The two surfaces may be amalgamated at Culver, or they may be so closely spaced that they were not resolved by our sampling.

The maximum flooding surface of Sequence 3 lies in the lower part of a distinctive sequence of less many chalk couplets in the middle M. dixoni Zone.

Sequence 4 Orbirhynchia mantelliana band 2, named after the common occurrence of the brachiopod Orbirhynchia mantelliana (J. de C. Sowerby), is located at base of the Middle Cenomanian Cunningtoniceras inerme Zone. The sequence boundary of Sequence 4 lies at the base of the marly interval that characterises 0. mantelliana band 2.

Transgressive surface 4 is placed at the base of the Cast Bed of Price (1877), couplet Cl of Gale (1995), a dark silty marl which at the base contains an abundance of the small scallop Entolium orbiculare (J. Sowerby) and a diverse fauna including the belemnite Actinocamax primus Arkhangelsky.

The maximum flooding surface of Sequence 4 is the calcimetry break at the base of couplet C 17, which marks the base of the Grey Chalk.

Sequence 5 A distinctive many chalk overlying an omission surface (DSV) coincident with the base of the Acanthoceras jukesbrownei Zone defines the sequence boundary of Sequence 5. Approximately 2 m above this, a distinctive thin grey marl (base of couplet DI) marks the bottom of a sequence of silty and arenitic chalks containing numerous laminated structures that have been variously described as burrow fills or scours. This interval (couplets 131-138) is commonly referred to as Jukes-Browne Bed 7, the base of which constitutes the transgressive surface of Sequence 5. Pycnodonte vesiculare (Lamarck) oysters are common above this surface.

The maximum flooding surface of Sequence 5 is located close to the Middle-Upper Cenomanian boundary (=base of the Calycoceras guerangeri Zone).

Sequence 6 The base of the Plenus Marl (DS VI; sub-plenus erosion surface) marks the sequence boundary for Sequence 6. The base of Plenus Marl Bed 4 is the transgressive surface (DSVII) of Sequence 6.

The base the Ballard Cliff Chalk (Melbourn Rock) constitutes a major facies change from many to nodular and intraclastic chalks (DSVIII). The maximum flooding surface of Sequence 6 is placed at the base of the overlying Holywell Chalk (=Meads Marl 6 of Mortimore 1986).

The above sequence stratigraphy formed the basis for comparison with our new chemostratigraphic data. However, in the North Cantabrian and Basco-Cantabrian basins of northern Spain, an additional sequence of earliest Late Cenomanian age is recorded, the base of which falls at or near the base of the Eucalycoceras pentagonuml Calycoceras naviculare (=C. guerangeri of this study) Zone.

Analytical Methods

Bulk sediment (125) samples were collected at c. I m intervals. Chalks and marls were both sampled; no attempt was made to select particular lithologies. Sample spacing was reduced across critical stratigraphic intervals such as the base of the Middle Cenomanian and the Cenomanian-Turonian boundary because these have been interpreted previously as representing periods of rapid environmental change which should be reflected in the geochemical signatures. Samples were ground to fine powders in an agate ball mill, washed with deionised water to remove modern sea-salt contamination and dried at 60 deg C; subsamples were fused with LiBO, and the melts dissolved in 0.5 M HN03. Silica, Ti, Al, Mn, Ca and Zr were determined by inductively coupled plasma-atomic emission spectrometry (ICP-AES) at Kingston University. Analytical procedures and numerical data are presented elsewhere.

A complementary suite of 85 samples was collected for stable isotope analysis. Samples were cleaned using 100, 1-12_02 followed by acetone, and dried at 60 deg C. Powders were reacted with purified orthophosphoric acid at 90 deg C and analysed on-line using a VG Isocarb device and PRISM mass spectrometre at the University of Oxford. Data are reported relative to the VPDB (Vienna Pee Dee Belemnite) standard; further analytical details are provided by Jenkyns et al (1994). Data for an additional 20 mid-Cenomanian samples were taken from Paul et al (1994).

Previous workers have used acetic acid leaches in an attempt to isolate 'carbonate-associated' elements in pelagic carbonates. However, these leaches dissolve both carbonates and manganese oxyhydroxides, and cause desorption of Mn from mineral surfaces. Experiments by Murphy (1998) have shown that 1 M acetic acid dissolves 90-99% of the Mn in Cenomanian chalks, and whole-rock trends are essentially identical to those obtained from partial leaches. Manganese shows no clear correlation with either Ca or Al, and is plotted here as elemental and Mn/AI curves; Mn/Ca essentially follows Mn. Whole-rock Ca data are presented as $CaCO_3$ since low-Mg calcite is the only significant Ca-bearing phase. Silica, Ti and Zr covary with Al due to carbonate dilution effects, but when normalised to Al provide valuable information on mineralogical variation in the siliciclastic fraction.

Manganese Chemostratigraphy

The Mn profile for Culver Cliff displays a general long-term decrease from the base of the Cenomanian to the Turonian. This corresponds to declining Si, Ti, Al, Zr and bulk insoluble residue contents that indicate a declining siliciclastic supply during the Cenomanian transgression. This is consistent with Mn being controlled partly by the detrital flux. However, Mn is assumed to be predominantly a carbonate-bound element, so the partitioning of Mn in chalks must be at least a two-component system with Mn being associated with both clay minerals and carbonate (with secondary Mn-oxyhydroxides also being developed in weathered sections).

Superimposed on the long-term trend are seven short-term cycles with prominent Mn maxima that do not correlate with either Al or Ca. The maxima of most of these cycles occur around the maximum flooding surfaces defined by Robaszynski et aL (1998), while minima are close to, or coincident with, sequence boundaries. Lowstand systems tracts generally have relatively low Mn contents, transgressive systems tracts display sharply rising Mn values that reach maxima around maximum flooding surfaces, before declining again through highstand systems tracts.

In addition to the six Mn cycles corresponding to the six established sequences in the Anglo-Paris Basin, a further cycle occurs in the Upper Cenomanian. Based on the Mn stratigraphy at Culver, the sequence boundary for this sequence is placed at the Mn minimum in the Amphidonte bed, a marker interval in the C guerangeri Zone containing two or three marls and characterised by the presence of occasional Amphidonte and subordinate Pycnodonte oysters.

The Amphidonte bed is a basin-wide marker that can also be traced throughout NW Germany. The transgressive surface of Sequence Sa is placed at a level of sharply rising Mn values occurring approximately 4 m above the Amphidonte bed, in a prominent thin grey flaser marl seam that overlies an omission surface. The maximum flooding surface is placed at the Mn maximum on the smoothed profile, which occurs at the base of sequence of three very prominent 10 cm thick dark marl seams in the middle C guerangeri Zone, termed here the Monument Marls. The lowest of these (Monument Marl 1) has a clearly defined omission surface at its base.

The long-term decrease in Mn is eliminated in the Mn/Al ratio profile, but the trends are much less coherent. Nonetheless, major peaks and troughs identified on the Mn profile are also displayed by Mn/Al, with the exception of the large Mn peak associated with the Cenomanian-Turonian boundary. Here, the sharp rise in Mn at the base of the Plenus Marl is eliminated, and the Mn maximum is shifted into the Lower Turonian, so long-term trends in Mn/Al ratios appear to reflect the sequence stratigraphy better.

Carbon Stable-isotope Stratigraphy

Calcimetry and Variation in the Siliciclastic Fraction

The calcimetry profile, shows reasonable agreement with the sequence stratigraphic model that predicts low carbonate in lowstand systems tracts, rising values from transgressive surfaces through transgressive systems tracts, with maxima at maximum flooding surfaces that plateau through the overlying highstand systems tracts. This correspondence is to be expected because clay: carbonate ratios were one of the criteria used by Robaszynski et al. (1998) to define their sequence stratigraphic framework in basinal sequences. Deviations from the predicted relationships may be explained in part by the relatively course sampling resolution adopted in the chemostratigraphic study. As expected, the Al profile is close to a mirror image of Ca and corresponds well to the gamma-ray logs for Culver and a closely correlative succession in the Boulonnais, northern France.

Transgressive surfaces and maximum flooding surfaces are picked out by positive excursions in Ti/Al, whereas sequence boundaries generally exhibit low Ti/Al contents; Zr/AI shows similar trends. It is notable that in most cases peaks are defined by several samples (i.e. extend over more than I m) and consequently represent relatively long-term (perhaps 60-100 ka, based on the cyclostratigraphy) changes in sediment composition, rather than reflecting mineralogical and/or chemical variation associated with sediments overlying individual omission or erosion surfaces. Titanium resides principally in heavy minerals: ilmenite, rutile, anatase and sphene, whereas Zr is located predominantly in zircon. Where a basaltic volcanogenic source can be excluded, increased Ti/Al ratios in deep-sea sediments correlate with increased current energy and detrital supply. Chalk tephras have low Ti contents that contrast to elevated Ti in the detrital fraction, so high Ti/Al ratios in chalks are interpreted as indicating pulses of increased current energy and siliciclastic supply.

Silica occurs in a much wider range of sediment constituents than Ti and Zr, including detrital quartz, biogenic opal (and its diagenetic products: opal-CT, chalcedony, quartz), glauconite and clay minerals. Interpretation of the Si/Al profile is ambiguous, but high and declining Si/Al values from the base of the Cenomanian and high Si/Al ratios at transgressive surfaces and maximum flooding surfaces correlate with increased detrital quartz and/or glauconite contents. The correlation between Ti and Si peaks is consistent with at least some Ti residing in titanomagnetite inclusions in detrital quartz, like those identified from Coniacian-Campanian chalks, also on the Isle of Wight.

loA

Manganese Chemostratigraphy

General trends in the Mn chemostratigraphy show remarkably consistent trends with respect to the sequence stratigraphic model. One discrepancy relates to the location of the Sequence 4 sequence boundary which is apparently overlain by sediments with high (but falling) rather than low Mn values. However, clear Mn and calcimetry minima on the smoothed profiles both occur in the Cast Bed, the base of which was interpreted by Robaszynski et aL (1998) as representing a transgressive surface. Repositioning the sequence boundary of Sequence 4 at the base of the Cast Bed, an alternative favoured by some workers, would remove this discrepancy. However, this would place the Ti, Zr and Si maxima at the sequence boundary rather than at the transgressive surface. The possible relocation of this sequence boundary warrants further study. A second apparent anomaly relates to the lowstand systems tract of Sequence 6 which has a high Mn content. However, Mn/Al ratios remain low, indicating that high Mn values are probably caused predominantly by the marked increase in clay content that occurs at the base of the Plenus Marl.

The Mn stratigraphy at Culver records the existence of an early Late Cenomanian sequence that corresponds in biostratigraphical position (lower C. guerangeri Zone or its equivalents) with that described from northern Spain by Wilmsen (1997) and Grafe & Weidmann (1998). Robaszynski et al. (1998) did not find any sedimentological evidence for this sequence in the Anglo-Paris Basin, but Fayau Hardgrounds 1 and 2 developed in shallow-water sands in Sarthe, NW France are here interpreted as expressions of, respectively, (1) the sequence boundary and (2) the transgressive surface of this sequence. The chalk facies was probably deposited in water too deep to clearly reflect the relatively small sea-level changes responsible for the formation of the C. guerangeri Zone sequence.

Marine Geochemistry of Mn

Manganese and Carbon-isotope Stratigraphy

Increased mid-ocean ridge volume caused by a period of enhanced spreading is regarded as being the main cause of eustatic rise in sea level during the Late Cretaceous. Crustal production rates during the Cenomanian Turonian were at their highest for the last 100 Ma, and were associated with increased mid-ocean ridge volcanism and the formation of extensive ocean basaltic plateaux. The decline in seawater 17 Sr/"Sr ratios through the Cenomanian supports a substantial hydrothermal input into the oceans, but the maximum rate of sea-level change driven by this mechanism is assumed to be in the order of 10 m Ma-.

The six depositional sequences in the Cenomanian each represent c. I Ma, so changes in sea-floor spreading rate would have to occur more rapidly than previously thought to explain such short term sea-level fluctuation. There is no evidence of permanent polar ice caps during the mid Cretaceous making a glacio-eustatic mechanism for rapid sea-level change unlikely, although episodes of high-latitude ice accumulation and associated sea-level fall have been postulated. Manganese enriched waters and bottom sediments are produced by hydrothermal processes at mid-ocean ridges, so extensive ridge volcanism and ocean plateau formation during the Cenomanian and early Turonian may have increased the global marine Mn flux, while pulses in hydrothermal activity may have been partly responsible for the rising Mn contents that characterise Cenomanian transgressive systems tracts.

Manganese deposition reached a minimum in the early Turonian during the overall peak Cretaceous highstand. It may be speculated that low Mn also relates to elevated atmospheric and oceanic P02 associated with low pCO_2 values that followed the Cenomanian Turonian boundary event. Higher oxygen concentrations in marine waters would have led to increased oxidation of organic matter in the water column and deeper and weaker sediment redox boundaries, both of which would have decreased the effectiveness of Mn redox cycling and Mn-carbonate precipitation.

Idealised relationships between chemostratigraphy, depositional sequences and sea level are summarised. Manganese exhibits minima around sequence boundaries and through lowstands, rising values from the transgressive surfaces through transgressive sequence tracts, maxima around maximum flooding surfaces, and declining values through highstands. Remarkably similar relationships between Mn curves and sequence stratigraphy have been described in Toarcian-Oxfordian, Tithonian Berriasian and Coniacian Maastrichtian pelagic limestones elsewhere, but previous authors have not examined the behaviour of other elements. Carbonate contents increase in response to rising sea level, but display sharp decreases at sequence boundaries and minima in lowstands systems tracts. Values increase through each transgressive systems tract, reach a maximum around the maximum flooding surface, and then remain high and constant through the highstand systems tract. Silica and trace-element (Ti, Zr) aluminium ratios peak around transgressive surfaces and maximum flooding surfaces, indicating pulses of increased siliciclastic input which are not confined to single omission/ erosion surfaces. This study has demonstrated that when Mn is combined with calcimetry, Si, Ti and Zr data, elemental chemostratigraphy provides robust criteria for developing sequence stratigraphic models and identifying sea-level change in pelagic-hemipelagic carbonate successions. However, manganese enrichment in Cretaceous carbonate sediments does not relate directly to oceanic anoxia in these sequences, so models for the formation of Mn ore deposits that invoke such a relationship may require re-evaluation.

Isotopic Constraints on Subduction Polarity

The Kohistan terrane of northern Pakistan is a superb example of juvenile crust that formed by magmatic additions at an intra-oceanic convergent margin. It is particularly important to tectonicists because its formation and accretion history constrains the Cretaceous palaeogeography of the Neotethys, and to igneous geochemists because it provides a complete and well-exposed section of primitive arc crust. Almost no isotopic data has been previously reported for the Kohistan intraoceanic arc terrane, and several important aspects about its tectonic significance remain controversial, particularly how wide was the oceanic seaway that separated it from Asia, and what was Kohistan's subduction polarity? We report here new major and trace element and isotopic compositions for igneous rocks that formed during the evolution of Kohistan as an intra-oceanic arc, and use these data to (1) confirm that Kohistan represents an intra-oceanic arc; (2) demonstrate that, prior to collision with the Karakorum terrane, the Kohistan arc faced north over a south-dipping subduction zone; and (3) infer that the Kohistan intra-oceanic arc originated near the present location of the DUPAL isotopic anomaly, at near equatorial latitudes.

Geological Setting

The Kohistan terrane is located in northern Pakistan and is bounded to the north by the Shyok (or Northern) Suture and to the south by the Indus Suture. Sedimentary sequences indicate that formation of the intra-oceanic Kohistan arc began in Early Cretaceous time, consistent with Ar/Ar and Rb/Sr ages indicating that the bulk of the igneous infrastructure formed between 110 and 90 Ma. The inference that Kohistan originated as an intraoceanic are results from its mafic bulk composition. presence of pillowed lavas and marine sediments, low-K nature of associated felsic rocks, absence of continental crust or detritus, and separation of the inferred intra-oceanic arc of Kohistan from continental crust to the north and south by unequivocal suture zones. Kohistan and Ladakh comprise correlative parts of this Cretaceous intraoceanic arc separated by the late Cenozoic Nanga Parbat Haramosh uplift. The intraoceanic phase of Kohistan lasted until sometime between 102 and 85 Ma, when Kohistan collided with the Karakorum (Asia) across the Shyok Suture. From this time until collision with India about 50 Ma ago, Kohistan existed as an Andean-type margin. This report focuses on the intra-oceanic phase of Kohistan.

The intra-oceanic arc crust of Kohistan consists of five principal units, extending from the Indus Suture in the south to the Shyok Suture in the north: (1) basal ultramafic mafic cumulates (Jijal Ultramafics); (2) Kamila Amphibolite; (3) Chilas Complex of mafic to intermediate plutonic rocks; (4) early bimodal suite of intrusive rocks and Gilgit gneisses; and (5) Chalt Volcanics. The early bimodal suite is a small part of the region. The bulk of the Kohistan batholith formed after collision of Kohistan and Karakorum, during the Andeanmargin phase of Kohistan, and is of no concern to this report. We report geochemical and isotopic data on Kamila amphibolites, Chilas Complex intrusive rocks, and Chalt Volcanics.

The Kamila Amphibolite belt is a composite mass dominated by amphibolite-facies meta-plutonic and metavolcanic rocks. Hornblende Ar Ar cooling ages of about 80 Ma indicate that regional amphibolite facies conditions existed prior to this time. Two types of amphibolites are identified in the field: (1) fine- to medium grained amphibolites, either homogeneous or banded and (2) homogeneous, medium-to coarse-grained amphibolites. The former are thought to be metamorphosed mafic to intermediate volcanics, whereas the latter are interpreted as metamorphosed gabbros and diorites. Our study was concerned with the metavolcanic amphibolites; we report no data for the second group. The Kamila metavolcanic amphibolites are subdivided into HFSE-enriched and -depleted groups, and are referred to as ‘E-type’ and iD-type` metavolcanic amphibolites by Khan et al. (1993). E-type amphibolites have MORB-like major and trace element compositions and do not show spikes in normalised trace element patterns characteristic of subduction-related igneous rocks. In addition they show REE patterns with modest depletions in LREE. In contrast, D-type amphibolites show normalised trace element patterns consistent with the participation of a `subduction component' (Pearce 1984). Kamila amphibolites derived from metamorphosed plutonic rocks show a very strong `subduction component', with high Ba/Nb. Kamila metaplutonic amphibolites are compositionally very similar to gabbronorites of the Chilas Complex. Khan et al. (1993) concluded that all three varieties of Kamila amphibolites were derived from a mantle source that was enriched in incompatible trace elements.

The Chilas Complex is more than 300 km long and up to 40 km wide, defining the spine of the Kohistan terrane. The Chilas Complex is subdivided into 'main gabbronorites' and a 'UMA association' by Khan et al. (1989). Gabbronorites comprise the vast bulk of the Chilas Complex; the distinctive cumulate ultramafic, mafic, and anorthositic rocks of the UMA association formed later. Gabbronorites define a calc-alkaline suite that range from about 51 to 59o/0 SiO_2, comparable to island-arc non-cumulates of Beard (1986). They show the distinctive chemical characteristics of convergent margin magmas, especially strongly positive spikes for Ba, K, and Sr, and negative Nb anomalies. They have modest LREE enrichments ($(Ce/Yb)_n$=2-3.5), with REE contents of about 10 x chondritic abundances. The rocks cooled and equilibrated under P-T conditions of the pyroxene granulite facies, estimated at 750-850 deg C and 5 6.5 kbar. In spite of the close temporal and spatial affinities of the two principal Chilas Complex components, significant differences in mineral compositions exist between gabbronorites (Olivine absent, magnetite-ilmenite, An_{40}- An_{64}) and UMA (Olivine abundant, Cr-spinel, An_{83}- An_{99}). The Chilas Complex was emplaced at c. 100 Ma into high levels of the deformed arc. For this report, we accept the Chilas Complex as developed during the intra-oceanic phase of Kohistan arc evolution, although Treloar et al. (1996) conclude that the Chilas Complex was emplaced after Kohistan collided with the Karakorum.

The Chalt Volcanics occupy a 330 km x 30 km belt along northernmost Kohistan. The upper parts of these volcanics are interbedded with the Yasin sediments of Aptian Albian age. Abundant pillowed lavas demonstrate that some if not all of these lavas erupted subaqueously. The Yasin and Chalt are strongly deformed and modestly metamorphosed, from lower greenschist facies in the west to lower amphibolite facies in the east. A significant compositional difference is observed along the strike of the belt, with a high-Mg suite encountered in the Hunza Valley (where the present suite was obtained) and a low to intermediate Mg calc-alkaline sequence found to the west. Boninite is common in the Hunza Valley section along with some tholeiitic basalts and abundant felsic lavas.

Analytical Techniques

Major- and trace-element analyses were obtained using standard XRF spectrometry techniques. The samples of Kamila Amphibolite and the Chilas Complex were analysed at the University of Oklahoma, using a Rigaku XRF, while those from the Chalt Volcanics were analysed at the University of Leicester, using Philips PW 1450 automatic XRF. Similar analytical techniques and standards were used at the two laboratories. Major elements were determined from glass fusion beads (made using a lithium tetraborate/ lithium metaborate flux; rock:flux ratio 1:5) analysed by using a Rh anode X-ray tube. Trace element concentrations were determined on pressed-powder pellets made using an organic binder (Moviol). Ni, Zr, Nb, Rb, Sr, Y determinations were made using a Rh anode X-ray tube with mass absorption corrections applied using the intensities of the Rh K alpha Compton scatter peak. Determinations of Cr and Ba used a W anode X-ray tube with mass absorption corrections applied using the intensities of the W L alpha Rayleigh scatter peak and Fe Ku to cross the Fe-absorption edge after the technique of Nesbitt et al. (1976). Detections limits (2Ó) are approximately I ppm for Cr, Rb, Sr, Zr, Nb and Y, and 2 ppm for Ni, Cr and Ba.

Concentrations of Nd, Sm. and Pb were determined by isotope dilution at UTD. Standard isotope dilution and cation exchange techniques were used to separate Sr. and LREE (Nd & Sm). A procedure modified after Richard et al. (1976) was used to separate Nd from Sm for isotope composition runs. The Pb separation procedure is reported in Manton (1988), with an initial anionic column separation. Total processing blanks were: $<$ 140 pg Pb; $<$700 pg Nd and $<$ 1 ng Sr. Isotopic compositions were determined with a Finnigan MAT 261 multi-collector thermal ionisation mass spectrometre; Nd was analysed in the dynamic multicollection mode whereas Sr and Pb were analysed in the static multicollection mode. Sr isotopic compositions were fractionation-corrected to $^{86}Sr/^{88}Sr=0.1194$ and adjusted relative to

E&A SrCO3 $^{87}Sr/^{86}Sr$=0.70800. Nd isotopic compositions were fractionation-corrected to $^{146}Nd/^{144}Nd$=0.7219 and ENd was calculated using the aNd values of Pier et ul (1989) for BCR-1 and UCSD standards. The standard values with reproducibility during the year in which isotopic analyses were carried out (I standard deviation, with digits corresponding to the last decimal places reported) were: E&A SrCO3 $^{87}Sr/^{86}Sr$=0.708020 (35; n=32), UCSD $^{143}Nd/^{144}Nd$=0.511848 (14; n=26), and BCR-1 $^{143}Nd/^{144}Nd$=0.512620 (8; n=8); NBS 981 (with 0.15% per amu Pb fractionation correction) ="*Pb/2')4Pb= 16.943, $^{207}Pb/^{204}Pb$=15.495. $^{208}Pb/^{204}Pb$=36.736 for 16 analyses.

Sr and Nd isotopic compositions listed are reported both as measured following fractionation correction ($^{143}Nd/^{144}Nd$) and, in the case of Sr, adjustment to a value of 0.70800 for E&A SrCO3 $^{87}Sr/^{86}Sr$, as well as initial isotopic compositions corrected for 120 Ma of radiogenic growth using values of $^{147}Sm/^{144}Nd$ and $^{87}Rb/^{86}Sr$ from element abundances listed. Initial isotopic compositions are reported either as $^{87}Sr/^{86}Sr$ or as Nd 120". Pb isotopic compositions are recalculated using the A notation of Hart (1988), which specifies the extent to which =$^{207}Pb/^{204}Pb$ and $^{208}Pb/^{204}Pb$ are distinct from a reference line defined by oceanic basalts from the northern hemisphere. This notation is used because it is an effective way to identify igneous rocks derived from certain mantle reservoirs such as the DUPAL source. Because U and Th concentrations have not been determined, the Pb isotopic compositions could not be recalculated as initial ratios. However, the isotopic composition of Pb in the samples may still be a useful isotopic fingerprint for the following reasons: (I) relatively little isotopic growth is expected to have occurred; and (2) relatively modest fractionation of U, Th, and Pb is expected during generation of mafic melts from the mantle. The Pb isotopic characteristics of Kohistan samples are similar to the DUPAL mantle reservoir, and this is not likely to be a result of post-eruption radiogenic growth.

The investigated Kamila samples are all mafic, with a restricted range in silica (48-52%) but with a much wider range in incompatible elements such as K. The subdivision of the Kamila metavolcanic amphibolites into E-type (relatively high Ti and other HFSE such as Zr, Y, and Nd; relatively low K_2O, Rb, and Sr) and D-type is supported by trace element ratios (E-type has high K/Rb=660 1330 and low Sr/Nd=8 13 characteristic of MORB and ocean island basalt (OIB), whereas D-type has low K/Rb= 330470 and high Sr/Nd = 18 46, similar to that of modern convergent margin volcanic suites). All of these features are consistent with E-type Kamila metavolcanic amphibolites having a MORB-like basaltic precursor, whereas D-type Kamila metavolcanic amphibolites have clear characteristics of supra-subduction zone igneous rocks. We will show that Kamila D-type metavolcanics are similar in many respects to the rocks of the Chilas plutonic complex.

The isotopic data also distinguish between the two types of Kamila metavolcanic amphibolites. E-type samples have more radiogenic Nd (õNd_{120} Ma=+7.0 to +7.4) and less radiogenic initial Sr ($^{87}Sr/^{86}Sr$=0.7036-0.7039) than the three D-type samples (õNd_{120} Ma = +2.9 to +4.7; $^{87}Sr/^{86}SrSr$=0.7041-0.7047). E-type and D-type amphibolites plot in distinct locations in a plot of Sr-Nd isotopes. The Pb isotopic compositions of E- and D-type metavolcanics are similar. Kamila E-type amphibolites have the most radiogenic Nd and least radiogenic Sr and $^{206}Pb/^{204}Pb$ of all the Kohistan samples that we have analysed. Kamila D-type metavolcanic amphibolites are similar to Chilas samples in having relatively low õNd_{120} Ma and high $^{87}Sr/^{86}Sr$, although only one of the D-type Kamila samples (A125) has a similar Pb isotopic composition. A puzzling feature is that Kamila amphibolites have homogeneous and relatively non-radiogenic $^{206}Pb/^{204}Pb$ and $^{208}Pb/^{204}Pb$, but have variable $^{207}Pb/^{204}Pb$.

Of the four Chilas samples that were analysed, two are from the Main Gabbronorites and two are from the UMA. UMA samples are cumulates, so it is difficult to infer much regarding their petrotectonic setting from their chemical compositions. The gabbronorite samples on the other hand may approximate melt compositions; they are mafic to intermediate and belong to a low-K to moderate K suite. Gabbronorites have high K/Rb and Sr/Nd; we can infer from their Sm and Nd abundances that these were moderately enriched in LREE. The isotopic composition of the four Chilas samples is very homogeneous, with $^{87}Sr/^{86}Sri$=0.70396 - 0.70414, õ$Nd_{(120)}$=+2.8 to +4.0, and very homogenous Pb. Along with one sample of D-type Kamila amphibolite (A125), the Chilas complex defines the relatively 'enriched' endmember of the Kohistan intraoceanic arc terrane. The Chilas Complex samples and some D-type Kamila amphibolites extend towards an EMI isotopic reservoir in Sr-Nd isotopic space, but extend towards EMII in Pb isotopic space.

The Chalt Volcanics define a compositionally heterogeneous group. They include low-K and medium-K rocks of basaltic, boninitic, and rhyodacitic composition. The suite is characterised by very low TiO_2 and Zr contents (<0.45% and <80 ppm, respectively). Chalt boninites and low-Ti basalts may contain very high contents of MgO (up to 12.5%), Ni (up to 400 ppm), and Cr (up to 1500 ppm). Boninitic samples (PL28, N138, N161, and IK580) have 10 to 12.5% CaO, and must therefore be classified as 'High Ca boninites' according to Crawford et al. (1989). Basalts and boninites have very low abundances of Sr (< 100 ppm), Ba (<40 ppm), and Nd (<8 ppm).

Samples of the Chat Volcanics that we analysed have homogeneous isotopic compositions of Nd ($\tilde{o}Nd_{(120\ Ma)}$= +6.3 to +7.3) and Pb (6 $^{206}Pb/^{204}Pb$=18.1 18.4 Ä ^{208}Pb=44-58; Ä ^{207}Pb=5 10). The 8 samples of Chalt boninite, basalt, and rhyodacite fall between Kamila and Chilas suites on plots of Pb isotopic compositions. The Chalt Volcanics define a much larger range on a plot of Sr v. Nd isotopic compositions, reflecting the larger range in H= $^{87}Sr/^{86}Sri$ (0.7040 0.7055). The Chalt Volcanics are metamorphosed submarine lavas, probably reflecting partial re-equilibration accompanying submarine hydrothermal metamorphism. Such metamorphism greatly affects Sr, but not Nd or Pb isotopic compositions. If so, the Chalt before being metamorphosed may have had as restricted a range of Sr isotopic compositions as it now has Nd and Pb isotopic compositions. The petrologic diversity of the Chalt Volcanics can be explained by a fractional crystallisation process from a parental magma similar to the Chalt boninites or basalts. Nd and Pb isotopic homogeneity for the large range of SiO^sub 2^contents in Chalt Volcanics supports this model, although this requires that the range in Sr isotopic composition reflects alteration. Alternatively, the isotopic data are consistent with a model of anatexis whereby Chalt boninites or basalts were melted to generate felsic lavas. It is also noteworthy that the Sr and Nd isotopic composition of the 102 +/- 12 Ma Matum Das tonalite, which crops out along the Hunza River in the belt of Chalt Volcanics, has Sr and Nd isotopic compositions that indicate close petrogenetic affinities with the Chalt.

The range of $^{87}Sr/^{86}Sr$ for igneous rocks of the intra-oceanic Kohistan arc (0.7036-0.7056; mean=0.7043) is similar to that for initial ratios for correlative units from Ladakh, including the Dras Volcanics (c. 0.7035), Kargil mafic cumulates (0.704-0.7053), and Ladakh intrusives (0.7034-0.7048). The data presented here allow us to address three questions about the Kohistan intra-oceanic arc system. These are: (1) polarity of the intra-oceanic Kohistan arc; (2) nature of the source region of Kohistan arc magmas; (3) palaeogeographic setting of the Kohistan intra-oceanic arc. These items are discussed further below.

Polarity of the Intra-oceanic Kohistan Arc

The intra-oceanic Kohistan-Ladakh arc formed over a subduction zone which dipped beneath it either to the south or to the north. From the time it was first recognised as an intraoceanic arc complex, investigators have preferred the interpretation that the dip was to the north. We are not convinced by the evidence for a northerly dip direction for the Kohistan subduction zone prior to collision with the Karakorum. Regardless of the orientation of the subduction zone beneath Kohistan, another northdipping subduction zone beneath Karakorum is required to generate c. 95 Ma granodioritic rocks of the Karakorum batholith. Collision between Kohistan and Karakorum to form the Shyok Suture is constrained to have occurred sometime in the early Late Cretaceous, 85 102 Ma ago. Pudsey (1986) argued that this suture represented a collapsed back-arc basin in part to accommodate the lack of evidence in the suture melange for subduction of a large oceanic tract.

Our understanding of the distribution of igneous rocks and their trace element and isotopic compositions in Kohistan presents another perspective for constraining subduction polarity of the intra-oceanic Kohistan arc. We conclude that the chemical and isotopic composition of Kohistan igneous rocks, particularly the Chalt and Kamila volcanic sequences, formed over a south-dipping subduction zone. We appreciate that our interpretation is not exclusive: for example, Treloar et al. (1996) infer that the Kamila provides substrate to Chalt cover. However, we find that the geochemical and isotopic data are more consistent with Kohistan sequences reflecting changes across an approximately contemporaneous arc system. There are two lines of evidence to support this. The Chalt Volcanics (especially their boninitic component) are interpreted to have formed in a forearc setting.

There are several models for why boninites form in forearcs, including subduction initiation, spreading-centre subduction, and interaction of the elevated thermal regime associated with a back-arc basin spreading centre superimposed on the high water flux of a forearc. Discussion continues regarding which of these models is most appropriate but the consensus that boninites form in fore-arc settings remains. The interbedded submarine boninites, low-Ti tholeiites and felsic rocks of the Chalt Volcanics are reminiscent of the Izu Bonin Mariana fore arc as is the rapid transition from volcanism to carbonate-dominated pelagic and hemipelagic sedimentation represented by the Chalt-Yasin succession. The MORB-like affinities of Kamila E-type metavolcanic amphibolites and their intercalation with D-type amphibolites which have a strong subduction component is most consistent with the interpretation that these formed in a back-arc basin. MORB-like magmas are only found at convergent margins in a back-arc basin setting, and sequences containing alternating or interfingering MORB- and arc-tholeiites are characteristic of modern back-arc basins such as the Mariana Trough and Lau Basin.

The inference that the 'subduction component` becomes less important from the Cha.lt Volcanics in the north to the Kamila Amphibolites in the south is also supported by an overall increase in the incompatible HFS elements Ti and Zr in the same direction. Abundances of HFS elements in an intraoceanic convergent margin increase from the forearc to the back-arc, which reflects an overall decrease in the degree of melting or degree of source region depletion that gives a clear facing direction for the arc. The HFSE evidence for Kohistan is very clear; considering only rocks with 48-59% SiO_2 the Chalt Volcanics contain <0.3% TiO_2 and <30 ppm Zr; Chilas Complex samples contain 0.84 and 1.24% TiO_2 and 71-133 ppm Zr; Kamila samples contain as much as 2.25% TiO_2 and 149 ppm Zr.

This ocean closed as a result of two subduction zones one dipping north beneath Karakorum and the other dipping south beneath Kohistan. This tectonic situation has a modern analogue in the Molucca Sea, south of the Philippines. The Molucca Sea is closing as a result of west-dipping subduction beneath the Sangihe Arc to the west and east-dipping subduction beneath the Halmahera Arc to the east.

There is no evidence for a hiatus in India's northward motion during the Late Cretaceous, and continuous northward motion requires a new subduction zone to form immediately after Kohistan and Karakorum collided. This new subduction zone dipped north on the south side of Kohistan. This reversal of subduction polarity beneath Kohistan during the Late Cretaceous is analogous to the situation inferred for the Solomon Arc during the Late Tertiary. Thickened crust of the Ontong Java Plateau arrived at the Vitiaz Trench on the north side of the Solomon Arc during the Miocene. This stopped south-dipping subduction and led to the formation of a new, north-dipping subduction zone southwest of the Solomons.

Our model satisfies all tectonic constraints, including: (1) India began moving north during the magnetic quiet zone, at approximately 120 Ma, and continued to move rapidly north until collision with Asia about 50 Ma ago. The northward movement of India reflects the development of subduction zones to the north at about 120 Ma. (2) North-dipping subduction is required beneath Karakorum to generate the Karakorum batholith. (3) Following the formation of the Shyok suture, the subduction zone associated with the northward motion of India dipped north on the south side of accreted Kohistan. This explanation is also consistent with the presence of S-dipping structures in northern Kohistan, which may have formed in association with a S-dipping subduction zone.

Nature of the Source Region of Kohistan Intra-oceanic Arc Magmas

One of the important controversies surrounding convergent margin magma-genesis concerns the relative contributions provided by mantle as opposed to subducted crust and sediments. Understanding this chemical budget is important not only for understanding arc magma-genesis, but also for understanding a wide range of other first-order questions about the earth, including whether or not the Earth's mantle and crust are still differentiating, whether the continents are growing or shrinking, and how the sources of hot-spot mantle formed. It is widely acknowledged that a hydrous subduction zone environment may be responsible for fractionating elements that are not fractionated in anhydrous magmagenetic environments such as hot spots or mid-ocean ridges. The most spectacular of these incompatible element fractionations in arcs is the relatively high abundance of fluid-mobile alkali metals and alkaline earths (K, Rb, Ba, Sr) relative to the abundance of less mobile high field strength elements (e.g., Ti, Zr, Y, Nb, Ta). Having acknowledged these fractionations and the fact that some elements are clearly derived from subducted components such as ^{10}Be, the question remains as to the relative importance of contributions of other incompatible elements that are extracted from the mantle compared to those that are recycled from subducted crust and sediments.

We argue that the distinctive isotopic signature of Indian Ocean mantle is readily recognised in the composition of Kohistan igneous rocks, and that this signature overwhelms any signature from subducted sediment. Hart (1984) recognised a zone of enriched mantle underlying equatorial oceanic regions which he called DUPAL. DUPAL mantle generates lavas with distinctive isotopic characteristics, including relatively low eNd and $^{206}Pb/^{204}Pb$, and relatively high $^{87}Sr/^{86}Sr$ and Delta8/4. Regions underlain by DUPAL mantle are presently far to the south of Kohistan, but DUPAL isotopic characteristics are readily identified in all igneous rocks from the intraoceanic phase of the Kohistan arc.

Another perspective is gained from comparison of Kohistan igneous rocks with Deccan traps. The Deccan Traps are flood basalts that erupted about 65 Ma when India passed over the Reunion hotspot. Most of the Deccan basalts have been significantly contaminated by continental lithosphere, but the Ambenali unit in the upper part of the sequence is thought to manifest predominantly asthenospheric isotopic compositions. The Ambenali basalts formed about 3500 km south of their present position and thus represent the composition of the asthenosphere beneath the subequatorial Indian Ocean basin at the end of Cretaceous time. They show DUPAL isotopic characteristics, especially high delta 8/4, and constitute an appropriate data set that can be used to examine the potential mantle controls on the isotopic composition of Kohistan igneous rocks.

Six samples of Ambenali basalts analysed by Lightfoot et al. (1990) have eNd of +4.3 to +7.5 and $^{87}Sr/^{86}Sr$ of 0.70390.7046, very similar to the range of Kohistan igneous rocks (õNd= +2.8 to +7.4 and $^{87}Sr/^{86}Sr$=0.7036-0.7055). Differences between the Sr and Nd isotopic systematics of Kohistan igneous rocks and Ambenali basalts consist of extension of the Kohistan suite to slightly lower õNd that nevertheless lie within the mantle array, and the higher $^{87}Sr/^{86}Sr$ of the Chalt Volcanics. The Pb isotopic compositions of Kohistan igneous rocks are also similar to those of Ambenali basalts. The Ambenali basalts have a somewhat lower range of $^{206}Pb/^{204}Pb$ (17.7-18.3) than Kohistan rocks (17.9 18.6), but are similar in lying above the northern hemisphere reference line (NHRL) in similarly distinctive fashions. On a plot of $^{207}Pb/^{204}Pb$ v. $^{206}Pb/^{204}Pb$, both define a trend that obliquely intersects NHRL at $^{206}Pb/^{204}Pb$ 17.6 and extends towards the EMII reservoir. On a plot of $^{208}Pb/^{204}pb$ v. $^{206}Pb/^{204}Pb$, both define a trend which lies above and parallel to NHRL. The Pb isotopic characteristics for both Kohistan and Ambenali are consistent with mixing between mantle reservoirs EMI and EMIT. Intriguingly, evidence of such mixing cannot be inferred from Sr-Nd systematics.

Palaeogeographic Setting of the Kohistan Intra-oceanic Arc

Our inferences regarding polarity of the intra-oceanic Kohistan arc and its formation in an oceanic realm dominated by DUPAL mantle sources lead us to propose a palaeogeographic model for Early Cretaceous time. It shows Kohistan lying within the DUPAL region, about 3000 km south of the Karakorum. The Shyok Sea, that part of Neotethys north of Kohistan, lies between Kohistan and Karakorum. India and Karakorum move northwards together until Kohistan collides with Karakorum, when a new, north dipping subduction zone is constructed on the south flank of Kohistan. The development of this new subduction zone may be manifested by the first stages in the development of the Oman subduction zone (obduction of the Semail ophiolite), about 90-95 Ma.

This model is subject to a number of uncertainties which affect the size of the seaway that lay between Kohistan and southern Asia. First of all, the location of the southern margin of Asia at this time is controversial. For example, we present the configuration of Scotese et at. (1988) which places it about 1 SoN, but Searle (1991) places it closer to 10 degN. Secondly, the location of Kohistan is not highly constrained. It must lie between India and Asia, and we infer from the isotopic data that it lay in the DUPAL realm, but where precisely? We have also assumed that the surface projection and location of the DUPAL anomaly has not changed significantly in the past 100 million years. These uncertainties indicate that we cannot use the isotopic data to quantitatively reconstruct Kohistan's paleolatitude in Cretaceous time; nevertheless it should be noted that the paleomagnetic study of Yoshida et al. (1996) place Kohistan slightly south of the Equator about 100 Ma ago.

Another way to look at these uncertainties is to try to minimise the size of the Shyok Sea. If we accept Searle's (1991) position for southern Asia at 10 degN and a location for Kohistan closer to the northern limits of the DUPAL anomaly, perhaps along the equator, the Shyok Seaway shrinks in width to perhaps 1000 km. There is nothing in our data that precludes alternate configurations such as this.

Geochemical and isotopic data reported here are most consistent with the model that the Kohistan intra-oceanic arc faced north, with forearc crust in the north (Chalt volcanics) and backarc crust in the south (Kamila amphibolites). The isotopic data indicate Kohistan igneous rocks formed by melting of a mantle source with a strong DUPAL isotopic signature. Such mantle sources exist in the oceanic realm far to the south of the present tectonic setting of Kohistan.

Chapter 2

Toxicological Testing and Environmental Chemicals

Current methods of risk assessment for some manufactured chemicals may not accurately predict the risks of exposure to humans and animals. In fact, we are beginning to realise that very low levels of exposure to some chemicals present in the environment disrupt the endocrine system, particularly during fetal development, at doses to which humans and animals are routinely exposed. Recent research on environmentally relevant exposures raises great concern since endocrine signals regulate the differentiation and growth of cells in fetuses. The endocrine system consists of cells that produce chemical signals or hormones that regulate the development and subsequent adult functioning of other cells in the body. As the fetus grows, hormones control the development of cells and thus determine the formation of all the body's organs.

Adult organs may never function properly if hormonal signals are disrupted at this crucial developmental stage. Even very small changes in levels of endogenous hormones, whether natural or experimentally induced, produce significant changes in organ function throughout the remainder of life. In short, organs, especially the reproductive organs and brain, are highly sensitivity to endocrine disruption during fetal development. For this reason, the major concern regarding endocrine disruptors is with exposure during critical times in fetal development when organs are forming. Exposure to endocrine disruptors at this stage may reduce or destroy reproductive capability and result in the loss of entire populations.

Not only has the issue of the unique vulnerability of the fetus been ignored in some recent reviews, but the term endocrine disruptor has been replaced by some authors with the term endocrine modulator. This change in terminology is intended to convey an entirely different meaning, namely, that these chemicals are causing a small adjustment in function. In fact, endocrine modulation typically occurs in adults, where alterations of the adult endocrine system, such as result from taking oral contraceptives, typically lead to transient - reversible - effects that disappear when exposure to the chemical ceases and the chemical is cleared from the body.

On the other hand, a chemical that interferes with the normal functioning of the fetal endocrine system and irreversibly alters the development of reproductive organs and the brain is appropriately described as an endocrine disruptor. We consider permanent effects of endocrine disruptors on physiological processes to be adverse. Even more disturbing, current toxicological testing for potentially harmful substances that are being released into the environment is designed to assess very high doses of chemicals, while recent findings show that many of the most harmful effects occur at doses so low they are rarely tested.

Dilution Not the Solution

Decades ago, the Great Lakes region was home to many predatory birds that fed on fish in the lakes. In the late 1960s, wildlife biologists observed that entire populations of birds, salmon, and trout were not successfully reproducing, and their reproductive organs were abnormal. Moreover, the children of women who ate these contaminated fish from the Great Lakes displayed developmental abnormalities and a decrease in I.Q. Scientists now know that these effects are related to exposure to endocrine-disrupting chemicals, chemicals that are ubiquitous in the environment. These chemicals are found in plastics, pesticides, detergents, cosmetics, fabrics, and building materials such as insulation and carpets, to mention just a few examples.

In addition, endocrine-disrupting chemicals are now found in lakes, rivers, oceans, and the air, where they contaminate fish and other animals. Currently, there are about 60 known endocrine-disrupting chemicals, but only a very small number of the approximately 80,000 manufactured chemicals in use today have been tested for endocrine-disrupting effects. New findings concerning these chemicals conflict with earlier toxicology studies that concluded that concentrations of endocrine-disrupting chemicals in the environment are safe. Recent legislation, including 1996 revisions of the Safe Drinking Water Act and regulations enforced by the U.S. Food and Drug Administrations (FDA) and the Environmental Protection Agency

(EPA), has resulted in a plan to reevaluate these chemicals. Legislation and regulations are being implemented even though many toxicologists have yet to accept the possibility of adverse effects from low-dose exposure to environmental chemicals. In fact, the toxicological research community remains deeply divided on the assumptions that underlie risk assessment. While some toxicologists and government regulators continue to defend the traditional model of risk assessment for all environmental contaminants, we offer a new model for endocrine-disrupting chemicals that challenges the traditional assumptions in toxicology, assumptions that have until recently governed our national regulatory processes.

False Assumptions

For many years toxicologists have studied adverse effects of chemicals that are hormonally active either through hormone receptors or other mechanisms. Yet in spite of our extensive knowledge of how certain hormones act naturally, toxicological studies of chemicals that mimic or block these hormones have not always taken this basic information into account. As a result, toxicology will now have to reevaluate the assumptions underlying current testing strategies for chemical toxicity. Almost all prior studies of hormonally active chemicals used study designs that were assumed to be appropriate for chemicals that do not directly damage genes - which endocrine disruptors are generally thought to be.

These methods are based on the assumption that high-dose testing predicts low-dose outcomes; in other words, information from testing a chemical at doses that may be millions of times higher than those encountered in the environment can be used to predict effects at environmental exposure levels.

This, in turn, is based on the assumption that a threshold exists below which no adverse effects are seen and that there is a monotonic dose-response cure; that is, as dose increases the response increases or stays the same, but the response never first increases then decreases. If either of these assumptions is incorrect with regard to an endocrine-disrupting chemical, then the acceptable safe dose derived from high-dose studies will be incorrect. This would be the case, for example, if the response to a chemical were to first increase at very small doses and then actually decrease at much higher doses - a non-monotonic dose-response curve. Recent research has shown that this, in fact, does occur.

It is inappropriate for regulatory agencies to base acceptable exposure levels on an assumption applied uniformly for all classes of chemicals when existing data contradicts that assumption in specific cases. Nonetheless, many toxicologists vigorously defend the concept of a measurable threshold for adverse effects and a simple, monotonic dose-response function for all chemicals. They do so because the assumptions appear reasonable on the surface and because, over many years, a variety of toxic and carcinogenic substances have been found to actually behave this way. However, in the case of endocrine-disrupting chemicals, recent research sheds doubt on the conventional wisdom that the "dose makes the poison."

No Safe Dose

Traditional toxicological studies aim largely at establishing a no-observed-adverse-effect level. To do so, researchers study very high doses that produced acute toxic effects - such as a decrease in body weight - in laboratory animals and then lower the dose to a point where no adverse effects are observed. There is seldom greater than a 50-fold difference between the lowest and highest doses in these studies. This type of testing, though apparently simple, is not without its problems. For example, the validity of this type of test is statistically limited by the number of animals required by regulatory agencies to be tested in each dose group. The smaller the group size, the more difficult it is to reliably detect an adverse effect.

Moreover, because the purpose of testing is to establish a safe level, doses below the apparent level at which there is no acute toxic effect are rarely tested; therefore, the assumption that very low doses are safe has rarely been investigated. Toxicologists Edward Calabrese and Linda Baldwin estimate that fewer than 2 percent of toxicological studies have been designed to assess the possibility that effects might occur at doses below those that do not produce acute toxicity. As a result, there has been a dramatic miscalculation of the doses of endocrine disruptors that can result in abnormal development in animals.

The Numbers Game

Some endocrine disruptors represent a unique class of toxicants that can act via the same mechanisms - for example by binding to a hormone receptor - as endogenous hormones. These chemicals provide a model to examine the validity of the two assumptions of a threshold dose and a monotonic dose-response function. We will review evidence from two experimental systems showing that current toxicological testing methods used to examine environmental chemicals are inappropriate for determining endocrine-disrupting effects at environmentally relevant doses. Additionally, these examples, along with other information we will present, provide an explanation for the dramatic miscalculation of the doses of endocrine disruptors that can result in abnormal development in animals.

In our research, we have shown how exposure to environmental chemicals that mimic the hormone estradiol affects the development of the prostate gland in male mice. The prostate develops from the embryonic tissue that forms a portion of the vagina in females, and like the vagina, it has estrogen receptors and responds to estrogen. To determine whether an environmental chemical has estrogenic activity, we first measured the effects of elevating the natural estrogen estradiol in male mouse fetuses via maternal treatment to levels at the high end of the normal physiological range. Next, we looked for similar effects from administering a dose of a chemical thought to mimic this very small increase in estradiol.

Specifically, we first increased the levels of estradiol in male mouse fetuses by an infinitesimal amount as measured in picograms, or trillionths of a gram. We increased the amount of free, biologically active estradiol in the blood of male mouse fetuses during the time that the prostate is forming - from the natural levels of 0.2 picograms per millilitre of serum in untreated males to 0.3 picograms per millilitre, a 50-percent increase of only one-tenth of a trillionth of a gram of estradiol per millilitre of serum. This very small increase in an already very low endogenous estradiol concentration resulted in significant changes in the prostate of treated males. There were 40 percent more glandular ducts and double the number of androgen receptors per cell. Androgen receptors determine the capacity of cells to respond to male sex hormones, and also mediate the effects of male sex hormones on the development and functioning of the prostate. Of great importance, the abnormal enlargement of the prostate in treated males was permanent and persisted into adulthood.

This experiment illustrates an important principle in developmental toxicology. Brief exposures to endocrine-disrupting chemicals that interfere with fetal hormones - even to low doses - during a critical period in development can result in irreversible lifetime effects, even though no trace of the chemical that caused this effect can be detected in the adult. Only if chemicals persist for a very long time after entering the body might some trace remain from exposure during fetal development. We next examined the effects on prostate development in mice of an estrogen-mimicking chemical called bisphenol A, which is the monomer used to manufacture polycarbonate plastic bottles, including baby bottles, the lining of food and beverage cans, dental sealants, and many other products. Free, biologically active bisphenol A is released into food and drinks stored in plastic containers or in cans and are also released from dental sealant placed on teeth.

In 1996, the plastics industry reported to the U.S. government that the dose of bisphenol A that produced no effect in their standard toxicological studies was 50 milligrams - or thousandths of a gram - per kilogram of body weight per day. We fed pregnant mice amounts of bisphenol A within the range being consumed per day by people through dietary and environmental exposure. These doses were thousands of times lower than those used in previous standard toxicological tests reported to have no effect. Pregnant females were fed bisphenol A at a dose of either 2 or 20 micrograms, or millionths of a gram, per kilogram of body weight per day for only seven days during the time that the prostate and other reproductive organs in male fetuses are forming.

We observed numerous effects in adult male offspring, including permanent enlargement of the prostate and a decrease in daily sperm production. In female offspring, we found an earlier onset of puberty. Other effects of bisphenol A on the breast and pituitary gland at doses below the previously reported nodose-effect level of 50 micrograms have also recently been reported in studies with rats. The 2-microgram dose of bisphenol A that significantly altered fetal development is 25,000 times below the 50-milligram dose that the plastics industry found to produce no adverse effects in animals tested by the traditional high-dose toxicological testing methods.

Margin of Error

How can there be such discrepancies between the effects of seemingly safe, high levels of endocrine-disrupting chemicals and drastically smaller doses of the same compounds? An experiment we conducted with the manufactured estrogenic chemical diethylstilbestrol (DES) provides an explanation for the huge error that can occur in estimating safe doses of endocrine-disrupting chemicals in traditional toxicological tests. DES is often used in toxicological studies as a control estrogen, since it is as potent as estradiol, but unlike estradiol, DES can be administered orally, which is the route of exposure for chemicals such as bisphenol A. When pregnant mice were fed very low doses of DES - 0.02, 0.2 or 2.0 micrograms per kilogram of body weight per day for seven days - using the same procedures as those for bisphenol A, we found a significant increase in prostate size in the adult male offspring relative to untreated males. Like estradiol and bisphenol A, then, DES administered during fetal development can, even at very low doses, produce significant adverse effects, in this case permanent enlargement of the prostate. However, with a higher dose of DES, 20 micrograms per kilogram, the prostate was actually smaller than in the low-dose animals and was no different from controls.

Finally, at a dose of 200 micrograms per kilogram, a dose 10,000-times higher than that which increased prostate size, the adult mouse prostates were significantly smaller than controls, after correcting for body weight. In addition, at this very high dose, the mice exhibited a 10-percent decrease in body weight relative to controls. By plotting these results on a graph, it's easy to see that administration of very low to very high doses of DES produce an inverted-U dose-response curve. That is, low doses stimulated inappropriate enlargement of the prostate while high doses inhibited normal prostate development, and males exposed to some intermediate doses did not differ significantly from controls. An inverted-U dose-response function was also found for adult prostate weight following prenatal treatment with different doses of estradiol. If low doses had not been investigated, the very small, 20-microgram per kilogram dose of DES would have been considered a no-adverse-effect-level dose.

In traditional studies of the effects of estrogens on the prostate, only high doses were used, and only the decrease in prostate weight was observed. However, other effects have been observed in adult mice that received DES in low doses during fetal life. A similar inverted-U dose-response curve was observed for a behavioural effect of the same prenatal doses of DES. Specifically, at low doses of DES, adult mice displayed increased territorial bahaviour, but at the highest dose, territorial bahaviour declined significantly. The current risk-assessment model - the basic design for toxicological tests - that assumes a simple monotonic relationship between dose and response cannot accommodate a data set in which response first increases and then decreases as the dose increases.

Circular Reasoning

In the traditional model, information from the high-dose study is used to predict effects at lower doses found in the environment, which are virtually never directly tested. This constitutes circular reasoning. That is, the assumptions drive the study design, and researchers simply don't examine low doses. Failure to directly study low-dose effects reinforces the misconception that there is, in fact, a safe dose for certain endocrine-disrupting chemicals. This has lead to false conclusions concerning safe doses of these substances. We have found permanent, and thus adverse, effects for every one of the endocrine-disrupting chemicals we tested at doses below the supposedly safe levels derived from results of very high-dose studies, including the plastic monomer bisphenol [A[18]] and two insecticides, DDT and methoxychlor. Methoxychlor is still in use in agricultural and other applications. It is likely that other endocrine-disrupting chemicals will also be found to produce adverse effects at doses below the established safe-intake level.

Depending on the dose used in a toxicological study, very different conclusions about the effects of an endocrine-disrupting chemical can be reached. This leads to the question of how doses are selected in traditional toxicological studies and how the data from these studies are used in risk assessment. In the traditional model, the highest dose is established at a level that produces acute toxicity, such as a 10-percent reduction in body weight. The middle dose is typically about five times lower than the highest dose, and the lowest dose used is typically 10 to 50 times lower than the highest dose. The lowest of these three doses is expected to be the no-observable-adverse-effect level. To account for several uncertainties inherent in extrapolating the results from animals to humans, and because individuals vary in their response to chemicals, this level is reduced by a factor of 100. This adjusted dose is supposedly safe and predicted not to increase risk from exposure to the chemical.

From these data, EPA establishes a level of safe exposure - a reference dose - and the FDA establishes an acceptable daily-intake dose. Is this method of calculating risk adequate to predict the possibility that much lower doses found in the environment might have effects mediated by hormonal response systems that operate at millions of times lower doses? In our study with DES, we found that the prostate of the developing fetus was permanently enlarged by a maternal dose 10,000 times lower than the 200-microgram-per-kilogram dose that caused an acute toxic effect, including decreased body weight and an inhibition of normal prostate development. We also found that a dose 10 times lower than the acutely toxic 200-microgram-per-kilogram dose did not significantly stimulate or inhibit prostate development relative to untreated males and did not significantly decrease body weight.

This dose would traditionally be considered the no-observed-adverse-effect level, and would be used as the basis for calculating safe exposure levels. Correcting 100-fold for uncertainties, the predicted safe dose of DES would still be 100-times higher than the maternal dose that permanently enlarged the prostate in male offspring. Is it reasonable to assume that other estrogenic chemicals will also show an inverted-U dose-response relationship similar to that for DES and estradiol? Unless all environmental estrogenic endocrine disruptors behave differently from our natural estrogen estradiol and the manufactured estrogenic chemical DES, we have to assume that in a traditional toxicology study design, the endocrine-disrupting effects at low doses - in many cases similar to amounts being consumed by people - could be missed. We propose that until endocrine-disrupting chemicals are tested using doses within the environmentally relevant range that people and wildlife are being exposed to, we cannot assume that these exposure levels are safe. Current testing procedures need to be revised, and chemicals in use today that are being released into the environment need to be tested for endocrine-disrupting effects at environmentally relevant doses.

No Threshold Dose

Not only do we find that very low doses of estrogenic chemicals can have significant adverse effects, we also maintain that there is no threshold, or level of exposure below which there is no effect. The no-threshold-dose hypothesis has been tested in sex-determination studies in turtles and in breast cancer studies in women. These studies have shown that there is no threshold below which there is no response to estrogenic chemicals. There may also be no threshold for a response to other endocrine disruptors that operate through different endogenous response mechanisms. The concept of a threshold for any potential toxicant is a theoretical concept because no matter how low the dose that gives a statistically significant response, it can always be argued that the threshold lies somewhere below that dose and was missed by the experiment. Also, even if a large number of animals are examined in a study, research results may show no significant difference between the control and the animals that received the dose.

This may suggest that there is a threshold. However, by further increasing the number of animals in these studies, a significant response may be revealed no matter how small the dose. In many cases, therefore, the question of whether a threshold exists cannot be resolved. However, in some circumstances, it is possible to show that a threshold dose will not be observed in an experiment involving administration of a chemical that mimics the activity of a natural hormone - such as estradiol - that is already present at levels above the threshold for responses. In this situation, whether or not there is a threshold for a particular response to the endogenous hormone is irrelevant, because responses to the hormone are being observed. For example, consider the situation in which the maternal ovaries and placenta are secreting estrogen that passes into the fetal circulation - which occurs in a human pregnancy - and estrogen causes irreversible developmental effects.

The threshold for an estrogenic effect is already exceeded by the animal's own estrogens before any dose of an estrogenic endocrine-disrupting chemical is administered to the animal. In rodents, for example, endogenous estradiol influences development of numerous tissues and already exceeds the threshold dose. Very small changes in circulation of estradiol produce lifetime differences in the functioning of organs. This natural estrogen thus causes irreversible effects. Any added dose, no matter how low, of an estrogenic chemical that increases the number of estrogen receptors activated by estrogen will permanently alter the course of development, and thus experimentally no threshold dose will be observed.

In some species such as turtles and lizards and in all crocodilians, temperature determines whether the animal develops testes or ovaries. Unlike mammals and birds, these animals possess no X,Y chromosomes that determine sex. Instead, the incubation temperature of the egg in which the embryo develops determines gonadal sex. For the red-eared slider turtle, for example, during embryonic life, low temperatures stimulate production of testes and produce males, while higher temperatures spur production of ovaries and produce females. At intermediate temperatures, different proportions of males and females are found, with the proportion of females increasing with temperature. The evidence is that the level of endogenous estrogen production in the egg is controlled by temperature and that estrogen mediates sex determination. If estradiol or other estrogenic chemicals are applied to the eggshell at temperatures that would normally lead to the development of testes, the undifferentiated embryonic gonads develop into ovaries. As temperature increases, lower doses of exogenous estrogen are required to produce ovaries in the embryos that would otherwise become males, because background levels of endogenous estradiol increase as temperature increases. Conversely, when eggs are incubated at the higher temperatures that normally produce all females, administration of an inhibitor of the enzyme aromatase - the critical enzyme involved in the synthesis of estradiol - results in a decrease in the proportion of embryos with ovaries. Instead, embryos develop testes.

Incubation of eggs at temperatures that produce a minority of females meets the criteria for testing the threshold-dose hypothesis. At these temperatures there is a measurable background incidence of sex reversal - formation of ovaries rather than testes. This is due to the elevated circulating level of endogenous estrogen in the eggs that occurs because of a slight increase in temperature over levels that result in 100-percent males. Analysis of three published estradiol dose-response data sets reveals no threshold dose. This means that any dose, no matter how low, will result in some response, and that the incidence of the response at any dose below the lowest one used in the experiment can be directly calculated. Thus, a dose that produces no effect does not exist. There can be no safe dose.

Numerous human studies have established that some women may develop breast cancer even if they are never exposed to any exogenous, endocrine-disrupting chemicals. Likewise, in animals, ovarian estrogens increase the incidence of breast cancer in a variety of experimental models, and the incidence of breast cancer is markedly lower in ovariectomised animals. Treatment of ovary-intact animals with any one of a wide variety of estrogens increases the incidence of breast cancer above that found in controls because there is an increase in estrogenic stimulation of the breast. Therefore, women, as a population, are already above the threshold dose for estrogen-induced disease, and exposure of breast tissue to estrogenic endocrine disruptors at any dose should increase the risk for breast cancer.

Reception Problems

The breast cancer example shows that exposure to endogenous hormones at concentrations within the normal range can, at some stages of life, induce adverse effects and contribute to the background prevalence of adverse effects within a population. Another example is that the incidence of testicular cancer in males has been related to fetal exposure to levels of natural estrogen. These findings lead to two conclusions. First, since the background incidence of adverse effects such as breast and testicular cancer is already high, there is no threshold dose for exogenous estrogenic chemicals to increase the likelihood of these adverse effects. This lack of a threshold dose is of profound importance. It means that every dose carries risk. There is no safe dose; there are only levels of risk that society deems acceptable.

Second, inverted-U dose-response curves are a consequence of normal physiological regulatory mechanisms under the control of endogenous hormones. Such non-monotonic, inverted-U curves are commonly encountered in endocrinology. One reason for a lower response at high doses is that high doses can decrease the numbers of hormone receptors in hormone-responsive cells while much lower doses of the hormone can produce an increase in receptors. Also, hormones usually bind to their corresponding receptors: estrogen to estrogen receptors, androgens to androgen receptors.

At high, toxicological doses, hormones may bind to inappropriate receptors. For example, at high doses, estradiol binds not just to estrogen receptors, but also to androgen receptors that normally bind to the masculinising hormone testosterone. This interferes with the effects of androgen in cells. During fetal life, this permanently interferes with normal masculinisation in males. Thus, at high doses, estrogenic chemicals interfere with a cell's ability to respond normally to other hormonal signals. Within the normal physiological range of hormonal activity, there is a delicate equilibrium that hormone-mimicking environmental chemicals disrupt. High or toxic doses of natural hormones or hormone-mimicking chemicals can produce effects that are dramatically different from low doses that disrupt this balancing act.

An important issue with regard to the concern over permanent effects of endocrine disruptors during fetal life is that effects in fetuses can be quite different from effects of the same chemicals in adults. This finding relates to whether the principle of homeostasis, the regulatory mechanisms that govern the maintenance of a constant internal environment, can be applied to embryos and fetuses based on studies of these same mechanisms in adults. While some regulatory mechanisms exist in the developing organism, homeostasis as described in adults does not. For example, administration of a high dose of estradiol to pregnant mice results in an increase in circulating testosterone in male fetuses, while high doses of the same hormone in adults result in a decrease in testosterone. In addition, female fetuses have high levels of pituitary gonadotropic hormones, which regulate estrogen levels in adults, together with high levels of circulating estrogen. This is clearly inconsistent with the homeostatic mechanisms of the adult female endocrine system in which estrogen exerts a negative feedback effect on the pituitary and suppresses the gonadotropic hormones. Therefore, studies of adults cannot predict how hormones and hormonally active chemicals will behave in the fetus.

No Safe Dose

Recent research has revealed new information about the way estrogenic endocrine-disrupting chemicals behave. This information seriously jeopardizes the validity of traditional toxicological studies. In fact, we can now say that there is a biologically plausible basis for the hypothesis that there is no safe or threshold dose of these chemicals. These issues, which have now been addressed for two classes of endocrine disruptors - estrogenic chemicals and dioxin-like chemicals - now need to be explored in many other experimental systems and for other classes of endocrine disruptors. As we obtain more information and a better understanding of how these previously unanticipated phenomena occur, the procedures for establishing socially acceptable exposure levels will likely change. If we accept that endocrine-disrupting chemicals do not show a response threshold in some circumstances, we will have to abandon the traditional concept that there are safe exposure levels that pose no increase in risk. We should acknowledge that there are risks involved in the use of products containing endocrine-disrupting chemicals and establish levels of risk that society deems acceptable.

The Biological and Chemical Reactivity of Zeolite-based Aluminosilicate Fibres

Environmental and/or occupational exposure to minerals, metals, and fibres can cause lung diseases that may develop years after exposure to the agents. The presence of toxic fibres such as asbestos in the environment plus the continuing development of new mineral or vitreous fibres requires a better understanding of the specific physical and chemical features of fibres/particles responsible for bioactivity. Toward that goal, we have tested aluminosilicate zeolites to establish biological and chemical structure-function correlations. Zeolites have known crystal structure, are subject to experimental manipulation, and can be synthesized and controlled to produce particles of selected size and shape. Naturally occurring zeolites include forms whose biological activity is reported to range from highly pathogenic (erionite) to essentially benign (mordenite). Thus, we used naturally occurring erionite and mordenite as well as an extensively studied synthetic zeolite based on faujasite (zeolite Y). Bioactivity was evaluated using lung macrophages of rat origin (cell line NR8383). Our obxive was to quantitatively determine the biological response upon interaction of the test particulates/fibres with lung macrophages and to evaluate the efficacy of surface iron on the zeolites to promote the Fenton reaction. The biological assessment included measurement of the reactive oxygen species by flow cytometry and chemiluminescence techniques upon phagocytosis of the minerals. The chemical assessment included measuring the hydroxyl radicals generated from hydrogen peroxide by iron bound to the zeolite particles and fibres (Fenton reaction). Chromatography as well as absorption spectroscopy were used to quantitate the hydroxyl radicals. We found that upon exposure to the same mass of a specific type of particulate, the oxidative burst increased with decreasing particle size, but remained relatively independent of zeolite composition. On the other hand, the Fenton reaction depended on the type of zeolite, suggesting that the surface structure of the zeolite plays an important role.

Epidemiologic data suggest that environmental and/or occupational exposure to minerals, metals, and fibres can cause lung disease (1-3). These diseases typically develop over many years after exposure to the agents. The most studied fibre is asbestos (1-3). Man-made mineral or vitreous fibres can also be bioactive, though their role in respiratory disease in humans is not yet well established and is an active area of study (4,5). There are more than 70 varieties of synthetic inorganic fibres, covering over 35,000 applications, with different physicochemical and morphological characteristics. These include insulation materials (glass wool, rock wool, slag wool), glass filaments and microfibres, and refractory ceramic fibres (4). As a consequence of the extensive applications of these fibres, a significant fraction of the population is exposed. Thus it is essential to understand the basis of toxicity of respirable fibres. In this study, we focused on developing a better understanding of the biological and chemical reactivity of aluminosilicate fibres and particles. Epidemiologic and experimental data have demonstrated that exposure to asbestos can induce pulmonary inflammation, fibrosis of the lower respiratory tract (asbestosis)(1-3,6) and is a risk factor for developing bronchiogenic carcinoma and mesothelioma. Numerous studies have been performed over the past 30 years to determine the mechanism(s) by which asbestos causes disease, and several hypotheses have been generated. Activation of macrophages by phagocytosis of the fibres results in the formation of reactive oxygen species (ROS) (1-3,6,9,10), where ROS is a collective term that includes radicals (superoxide anion, hydroxyl, peroxyl, and alkoxyl radicals), and hydrogen peroxide (H_2O_2). Iron bound to the fibres can generate hydroxyl radicals via the Fenton reaction, which can initiate lipid peroxidation, resulting in the production of intermediates that oxidise intracellular proteins and DNA (11,12).

In this study, we focused on separating the biological response (oxidative burst) and the chemical reactivity (Fenton reaction) using aluminosilicate crystals belonging to the zeolite family. There are several reasons we chose to examine zeolites as model systems. Even though the exposure of zeolites to the general population is quite limited, with possibly the exception of detergents, there is a particular zeolite called erionite that is highly toxic and causes mesothelioma (13,14). Several other zeolites have also been examined for their toxicity, but none so far has paralleled the effects of erionite. Erionite is a naturally occurring zeolite mineral that is far more carcinogenic than crocidolite asbestos (13,15). The framework of erionite is made up of interlocking tetrahedra of silicate and aluminate tetrahedra, is negatively charged, and can bind cations. The toxicity of natural erionite is commonly associated with the iron that accumulates on its surface after it is deposited within the respiratory epithelium (16,17). Mordenite is a natural zeolite with a chemical composition similar to erionite, as well as ion exchange abilities and to some extent a fibrous morphology. It is not reported to be carcinogenic and has been classified as being "slightly biologically active" (17,18).

In this study, we examined three zeolites, two of which are mineral zeolites, erionite and mordenite, and zeolite Y, a synthetic zeolite, extensively used as a catalyst. The biological response and surface chemical reactivity of the zeolites have been studied. For assessment of biological response, we examined the oxidative burst as a result of phagocytosis by rat pulmonary alveolar macrophage-derived cells (NR8383). We chose these cells because they can be propagated in vitro and exhibit a number of important macrophage characteristics, such as surface Fc receptors, interleukin-1 production and secretion, and oxidative burst response. Our previous studies dealt with phagocytosis of erionite by the NR8383 macrophages (20,21).

For determining chemical reactivity, we focused on the ability of the iron-exchanged forms of the zeolites to produce hydroxyl radicals from H_2O_2 (Fenton reaction). Both the intracellular and extracellular release of ROS by means of flow cytometry and chemiluminescence was studied. To distinguish cellular responses induced by fibre interactions with the plasma membrane from those induced by phagocytosis, we treated subsets of cells with the endocytosis inhibitor ammonium chloride. For surface chemical reactivity, we focused on the ability of the zeolites to promote Fenton chemistry. Total exchanged iron was quantified by elemental analysis, and surface iron was determined by a novel method utilising the ion-exchange of surface iron by a positively charged polymer. To quantify the production of hydroxyl radicals, we used DMSO as a trapping agent. Its decrease was monitored by reverse phase liquid chromatography (RPLC) and by the formation of methanesulfinic acid (MSA), a stable product of the reaction of DMSO with hydroxyl radicals by electronic spectroscopy. Conclusions regarding the dependence of the oxidative burst on particle size and zeolite composition and the role of the zeolite surface structure on the Fenton chemistry are presented.

Cell Culture

The rat pulmonary alveolar macrophage-derived cell line, NR8383, was obtained from R. J. Helmke. The cells were cultured in Hams F12 medium supplemented with 15% fetal bovine serum (Ó) at 37°C in a humidified atmosphere of 5% C[O_2] in 30-mL polystyrene culture flasks. Cell populations were expanded by weekly passage at a ratio of 1:2. We determined cell viability by trypan blue exclusion.

Zeolite Characterisation

Size Fractionation

Approximately 1 g of mordenite or erionite (Minerals Research, Clarkson, NY) was added to 140 mL of water in 150-mL beakers, and allowed to settle for 70 and 96 min, respectively (determined by fibre density). The supernatant was then removed, filtered under vacuum through a 0.80 pm pore size filter and dried in an oven at 70°C. Zeolite

Y obtained from Union Carbide (Danbury, CT) did not require size fractionation. We determined particle sizes of the fractionated materials by scanning electron microscopy (SEM). Diffraction patterns were collected using a Rigaku Geigerflex D/max2B diffractometre (Rigaku, The Woodlands, TX) using Ni-filtered CuK[α] radiation or a Bruker AXS D8 Advance diffractometre equipped with a rotating tilted sample holder. X-ray diffraction pattern peaks were search-matched using Jade 5 matching software (Materials Data, Inc., Livermore, CA).

Incorporation and Determination of Iron on Zeolite

We placed 1 g of ground zeolite in a 12 mL centrifuge tube, to which 10 mL of a 0.1 M NaCl solution was added. The suspension was sonicated for 15 min, shaken on a wrist shaker for 15 min, centrifuged, and the supernatant removed. This step was repeated twice. We deoxygenated solutions of [10^{-4}], [10^{-2}], 5 x [10^{-2}] and [10^{-1}] M iron sulfate heptahydrate (Ó) with bubbling [N_2] for 30 min and acidified them to pH 4 by adding concentrated hydrochloric acid (Fisher). Zeolite samples were added to the iron solutions and the suspensions shaken for 2 hr. After filtering, the zeolites were rinsed with deoxygenated water 6-10 times under [N_2] atmosphere. We analysed rinse waters by inductively coupled plasma-optical emission spectroscopy (ICP-OES) and a spectrophotometric method to ensure complete removal of adsorbed iron. Samples were dried overnight under a flow of nitrogen. ICP-OES experiments were done on the solutions obtained from lithium metaborate fusion of the zeolites and analysed for iron (emission lines at 238.204, 239.562, 259.939, and 234.349 nm), aluminium (396.153, 308.215, 394.401, and 237.313 nm) and silicon (251.611, 212.412, 288.158, 252.851, and 221.667 nm) using an Optima 4300DV instrument. Molybdenum was used as an internal standard at 5 ppm (emission lines at 202.03, 203.844, and 204.598 nm).

Spectrophotometric Determination of Solution Iron

This method has been described by Fritz and Schenk. Solutions of 5 mM phenanthroline, 0.3 M ascorbic acid (Aldrich), and 1 M sodium acetate were prepared in deionised water. The spectra were measured with a Shimadzu UV 265 or a Shimadzu UV-2501PC spectrophotometre. Standards were prepared with iron sulfate.

Polymer Exchange for Surface Iron Determination

We used the polymer Dab-4Br, first described by Daniels et al. and obtained by synthesis of 1,4-diazabicyclol [2.2.2] octane (Dabco) with Br[(C[H_2])$_4$]Br. Solutions of 0.5 g of Dab-4Br in 40 mL water were prepared, and 8 mL of this solution was added to 0.4 g of dried iron-exchanged zeolite samples in a 12-mL centrifuge tube. The mixture was placed on a wrist shaker for 20, 40, or 60 hr. We then centrifuged the solutions and analysed them for iron. Diffuse reflectance. UV-Vis diffuse reflectance spectra were acquired on a Perkin Elmer Lambda 900 UV-Vis-NIR spectrometre. We used Kubelka-Munk units. The reference sample was the unexchanged zeolite.

Cell-Particle Interactions

We obtained the fluorescent dye 5- (and 6-) carboxy-2',7'-dichlorodihydrofluorescin diacetate ([H_2]DCF-DA) from Molecular Probes. We divided 25 mg of [H_2]DCFDA dissolved in 5.0 mL of anhydrous DMSO (Aldrich) into aliquots and stored them at -4°C in a freezer/desiccator until use. NR8383 cells were pelleted by centrifugation (400 x g, 10 min) and resuspended in fresh Hams F12 medium (-2.5 x [10^4] cells/mL). Cells were treated with [H_2]DCF-DA (final concentration 20 [micro]M) and incubated for 20 min at 37°C in a humidified atmosphere of 5% C[O_2]. Zymosan A (Ó), size-fractionated erionite, mordenite, and zeolite Y were each suspended in medium (100 pg/mL) and sonicated for 20 min immediately before addition to cells.

After incubation, we treated 1-mL aliquots of cells (2.5 x [10^4]) with 1 mL of fibre suspension (100 [micro]g particles/mL) in culture medium or in ACAS buffer [127 mM NaCI, 3.8 mM KCl, 1.2 mM CaC[I_2], 5 mM glucose (Jenneile Chemical), 0.8 mM Mg[Cl_2], 1.2 mM KH_2PO_4, and 10 mM HEPES (Ó), pH 7.4]. For negative controls, we treated additional aliquots of cells with 1 mL of medium or buffer alone. To promote cell/fibre contact, cell suspensions were vortexed briefly (10 sec) then gently centrifuged (200 x g, 3 min). This cycle was performed three times, after which cells were analysed as described below. To distinguish cellular responses induced by fibre interactions with the plasma membrane from those induced by phagocytosis, we incubated subsets of cells for 1 hr in medium supplemented with the endocytosis inhibitor N[H_4]Cl(Mallinckrodt), at a final concentration of 20 mM. Cells were then washed and resuspended in fresh N[H_4]Cl-supplemented medium before treatment with particulates.

Flow cytometry. Flow cytometric analysis was performed on a Coulter Epics Elite flow cytometre equipped with a 15 mW air-cooled argon ion laser operating at 488 nm. Fluorescence emission was collected using a 525 nm band pass filter. We measured the uptake of particulates by viable cells (determined and gated from a bivariate histogram of right-angle light scatter versus forward-angle light scatter) by the increase in the right-angle scatter signal as displayed on a univariate histogram of right-angle scatter versus the number of events. We used the mean right-angle scatter channel number for data analysis, as well as the mean fluorescence intensity and percentage of cells exceeding values measured for untreated control populations.

Chemiluminescence assay: Luminol (Ó) was dissolved in DMSO at a concentration of 10 mM and stored in aliquots at -4°C. Just before use, luminol was thawed and diluted to 100 [micro]M in phosphate-buffered saline (PBS), with a resultant final concentration of 1%. Zymosan A (Ó), size-fractionated erionite, mordenite, and zeolite Y were each suspended in PBS at concentrations of 5.0, 1.0, and 0.2 mg/mL and sonicated for 20 min immediately before treatment of cells. NR8383 cells were pelleted by centrifugation (400 x g, 10 min) and resuspended in fresh Hams F 12 medium (1.0 x [10^6] cells/mL).

We performed luminescence reactions in 96-well black Optiplates (Packard, Meriden, CT). To each well we added 100 [micro]L of luminol (final concentration in total volume of well was 40 [micro]M), 50 [micro]L of fibre suspensions (250 [micro]g, 50 [micro]g or 10 [micro]/well), and 100 [micro]L of cell suspension (1 x [10^5] cells/well, added last to synchronize kinetics among reactions). All reactions were performed in duplicate. As controls, each experiment also included duplicate wells containing luminol alone, cells alone, or cells and luminol but no fibres. Plates were sealed and luminescence was immediately measured with a Top-Count Microplate Scintillation and Luminescence Counter (Packard). After this initial measurement, plates were centrifuged (400 x g, 1 min) to. promote fibre/cell contact. We subsequently performed luminescence measurements at 3-min intervals over 2 hr. Plates were kept in an incubator at 37°C between each measurement.

To distinguish cellular responses induced by fibre interactions with the plasma membrane from those induced by phagocytosis, we incubated subsets of cells for 1 hr in medium supplemented with 20 mM NH_4Cl before use in experiments, anJ all components of reactions were likewise supplemented to attain a final NH_4Cl concentration of 20 mM in reaction wells. We obtained corrected chemiluminescence curves by subtracting values generated by negative control reactions (cells plus luminol in the absence of fibres) from those generated under each experimental condition. We then calculated total luminescence values by integration of the area under each corrected chemiluminescence curge (Grams/32 program; Thermo Galactic, Salem, NH).

Transmission electron microscopy: To determine the effectiveness of the endocytosis inhibitor NH_4Cl, we incubated NR8383 cells for 1 hr in culture medium in the presence or absence of 20 mM NH_4Cl, then exposed cells to erionite or mordenite using the vortex/centrifugation method described above. Cells were fixed, embedded, sectioned, and imaged by transmission electron microscopy as previously described.

Hydroxyl Radical Measurements

Reverse-phase liquid chromatography: We used a Waters C18 Resolve1 10 [micro]m column (8 mm x 10 cm; Waters, Milford, MA) and a Shimadzu LC-10ATVP HPLC system coupled to an SPD-M10AVP diode array detector (Shimadzu). The injection volume was 20 [micro]L. The eluant consisted of 5% methanol in nanopure water, filtered through a Duropore filter and degassed the day of use. We used an isocratic flow of 1 mL/min. The sample loop was rinsed six times with the sample and flushed six times with eluant after injection. All samples were injected in triplicate. Four standards of DMSO/H_2O_2 solutions were run. Integration of the peak areas was performed using Grams/32 software.

We added 10 mL of a 0.384 mM H_2O_2 and 0.939 mM DMSO solution to the dried iron-exchanged zeolite, and the mixture was shaken for 30 min on a wrist shaker. The solution was then centrifuged, and a fraction of the supernatant was filtered and injected into the chromatographic system. Centrifugation was performed using a Beckman-Coulter Alegra 64R centrifuge at 25,000 rpm for 10 min. To determine the influence of any iron released from the zeolite into solution, 2 mL of the last rinse solution was added to 10 mL of the H_2O_2/DMSO mixture, shaken for 30 min, and examined by chromatography.

Spectrophotometric determination of mettanesulfinic acid formation: Babbs and Steiner (24,25) developed the spectrophotometric quantification of MSA, and we used a modified version of this method. The photooxidised species from the Fast Blue BB salt solution $([C_{17}][H_{18}]Cl[N_3][O_3])Zn[Cl_2]$ was removed by extraction with 1 mL chloroform. We added 0.3 mL of 1N HCl and 0.2 mL of 0.02 M Fast

Blue BB salt solution to 2.0 mL of sample (obtained after zeolite treatment with DMSO and H_2O_2). The tube was kept in the dark for 10 min. We added 1.5 mL of a solution of toluene-butanol 3:1, and then the mixture was vortexed for 120 sec and centrifuged. We then removed 1.0 mL of the organic phase, placed it in a second tube, and added 2.0 mL of butanol-saturated water. The mixture was vortexed 30 sec and centrifuged briefly. We removed 0.9 mL of the upper phase and placed it in the UV cuvette, along with 1.5 mL of the butanol-toluene 3:1 solvent. Spectra were recorded on a Shimadzu UV 265 or a Shimadzu UV2501PC spectrophotometre, from 500 to 330 nm. We used 2.4 mL of the 3:1 mixture of toluene-butanol as a blank. Standards were run by repeating this procedure with solutions of 0.01 mM to 0.15 mM of MSA (obtained from Lancaster Laboratories, Pelham, NH).

Physical Characteristics of Zeolites

The powder diffraction patterns compare well with previously published data on all three samples. For erionite, the peaks at 20 values of 12.2° and 38.0° match the characteristic peaks of calcium aluminium silicate hydrate ($[Ca_4][Al_8][Si_8][O_{32}]*8[H_2]O$), and the peaks at 12.2°,17.5°, and 55.9° most likely belong to sodium aluminium silicate hydrate ($Na_3Al_3Si_5O_{16}*6H_2O$). No extraneous phases were evident in the X-ray diffraction pattern of mordenite and zeolite Y.

Scanning electron micrographs of an erionite dispersion before and after fractionation show that the average length and width of the fibres has effectively decreased, as well as the size range. The unfractionated erionite sample includes long fibres, up to 45 [micro]m, with a mode (most frequent length) of approximately 10 [micro]m. Fractionation of this sample led to a population ranging from 0.7 [micro]m to 13 [micro]m, with a mode of about 3 [micro]m, the fibre morphology appearing intact.

Fibrous morphology on mordenite was not frequently observed. Depending on the geographic source of mining, mordenite may not be fibre-shaped. We observed that unfractionated mordenite particles range from 0.1 to 10 [micro]m, with a mode of approximately 3.7 [micro]m. Size fractionation of mordenite allowed discrimination against the largest particles, as well as particles smaller than 1 [micro]m. The most abundant size observable on the SEM is centred on the value of approximately 1 [micro]m. Zeolite Y appeared as octahedral crystals of a homogeneous size of slightly less than 1 lam.

Cell-Fibre Interactions

Morphologic characteristics of fibre-treated cells: Morphologic characteristics of cells determine their light-scattering properties. The amount of light refracted in a forward direction (forward angle light scatter) is proportional to cell size and the refractive index of the cytoplasm. In contrast, the amount of light reflected at right angle (side scatter) is largely determined by the intracellular morphology of the cell, such as differences in nuclear size or shape, cytoplasmic granules or vacuoles, or bound particles. By plotting the side-scatter versus forward-scatter signals of cells in suspension, it is possible to distinguish single viable cells from nonviable cells (based on differences in refractive index) and cell dusters (based on differences in cell volumes). To show whether cells had phagocytised zeolite particles, side scatter was plotted only for the living cell population. The mineral suspensions alone showed a small forward-scatter and a relatively wide side-scatter range but did not overlap with the living cell population.

Side scatter increased with time of exposure to particles, so samples were analysed exactly 20 min after exposure to particles. For zeolite Y, the increase was 25%. Erionite particles induced a large increase in side scatter, with an average increase of 58%. Mordenite induced the second largest increase, reaching a value of 41%. In all cases, smaller particles, obtained via fractionation or sonication, induced a smaller increase in side scatter than their corresponding larger counterparts.

$N[H_4]Cl$ was used as the endocytosis inhibitor to inhibit association of the macrophages with particulates. Cell viability, evaluated using trypan blue after 1 hr exposure to $N[H_4]Cl$-enriched medium, was unaltered as compared to a blank. Cells treated with $N[H_4]Cl$ and exposed to fractionated erionite, fractionated mordenite, and zeolite Y all showed relative increase in side scatter that was only about 75% of that observed in cells exposed to particles in the absence of inhibitor.

Oxidative burst. We measured the oxidative burst as both the mean fluorescence intensity generated by viable cell populations and the percentage of cells with fluorescent intensities above the background level of untreated controls. A large positive signal was reproducibly observed in the case of zymosan for three replications. A typical result was that the percentage of cells generating fluorescent signal in excess of untreated controls increased from 4% to 53% of the total viable cell population upon exposure to zymosan, and the mean fluorescence intensity increased approximately 13-fold. However, despite extensive efforts, flow cytometry did not exhibit significant increase in fluorescence after exposure of the cells to the three different minerals. We attempted various modifications, including use of a different fluorochrome (dihydrorhodamine123), variation of exposure time, and modification of the centrifugation/vortex protocol, but all failed to induce a detectable oxidative response from the cells.

For all three zeolites, the peak luminescence occurred between 20 and 25 min after exposure. These data demonstrate dose-dependent responses in all cases: the greatest luminescence signal was always associated with the highest mineral concentration. In addition, we observed that for exposure to the same mass of particulates, the chemiluminescence signal was inversely correlated with particle size for each category of zeolites. To distinguish cellular responses induced by fibre interactions with the plasma membrane from those induced by phagocytosis, we treated subsets of cells with the endocytosis inhibitor $N[H_4]Cl$. Chemiluminescence assays of responses of $N[H_4]cl$-treated cells after exposure to fibres demonstrated that inhibition of endocytosis dramatically attenuated oxidative burst intensity, with typically about 70% reduction in levels of chemiluminescence. Although absolute levels of luminescence were reduced, $N[H_4]cl$-treatment had little effect on the kinetics of fibre induced oxidative burst. These data suggest that fibre-induced ROS production is primarily dependent on fibre phagocytosis, rather than fibre-membrane interactions.

Electron Microscopy

Transmission electron microscopy of untreated and $N[H_4]Cl$-treated cells followed by exposure to erionite or mordenite. The micrographs suggest that, although $N[H_4]Cl$ treatment did not affect binding of fibres to the plasma membrane, it reduced the quantity of intracellular fibres from about eight per cell (untreated) to one to two per cell ($N[H_4]Cl$-treated) (based on a study of four micrographs).

Chemical Studies

Bulk and surface iron analysis. Mordenite, erionite, and zeolite Y had initial levels of total iron of 4.7 x $[10^3]$, 12.7 x $[10^3]$, and 0.35 x $[10^3]$ [micro]g iron/g of zeolite, respectively. Levels of total iron in erionite dropped from 12.7 x $[10^3]$ to 10.3 x $[10^3]$ [micro]g iron/g and for mordenite from 4.7 x $[10^3]$ to 3.6 x $[10^3]$ [micro]g iron/g after sodium ion exchange, indicating that the iron is either part of the framework or present as an impurity phase. The iron levels on as-synthesized zeolite Y were the lowest. An increase in concentration of iron in solution resulted in an increase of the total exchanged iron. As expected from exchange isotherms, this increase is not linear and tends to level off at higher concentrations. Surface areas of 86 [m^2]/g, 295 [m^2]/g, and 518 [m^2]/g were determined for mordenite, erionite, and zeolite Y, respectively. Mordenite has the lowest surface area because of structural defects that block the channels, and for this reason it is also the poorest ion-exchanger of the three zeolites.

The spectrophotometric method to determine surface iron content is based on back-exchange of iron by a charged cationic polymer, Dab-4Br. The released iron after 20, 40, and 60 hr were measured and found to level out after 40 hr. We therefore report the results as the levels of iron exchanged after 40 hr of ion-exchange. To confirm that the polymer was exchanged only at the zeolite surface, we monitored the X-ray photoelectron spectroscopy before and after exposure. These studies showed a decrease in the Fe/Si ratio; however, the iron signal did not fully disappear. We attribute this to signal from the bulk iron, since X-ray penetration depth is approximately 50 [Angstorm]. The small standard deviations for surface iron levels indicate good reproducibility of the method. Surface iron increases with the concentration of the initial iron solution, ranging from 1.4 [micro]g/g to 12.3 x [10^2] [micro]g/g for zeolite Y, from 5 [micro]g/g to 4.0 x [10^2] [micro]g/g for mordenite, and from 1.6 [micro]g/g to 3.3 x [10^2] [micro]g/g for erionite. Levels of surface iron in natural samples before any exchange are 30 [micro]g/g and 4 [micro]g/g for mordenite and eftonite, respectively, values that drop to undetectable levels after sodium exchange.

The large increase of surface iron on zeolite Y when exposed to iron sulfate solutions of 0.1 M was examined further. The pH of the initial solution of acidified iron sulfate was 3. X-ray diffraction patterns of the zeolite Y after exposure to these solutions confirmed a loss in crystallinity, whereas the structure was stable for both erionite and mordenite. The lower Si/Al ratio of zeolite Y makes it more prone to degradation. Consequently, the value of 12.3 [+ or -] 0.3 x [10^2] [micro]g of surface iron/g of zeolite Y represents iron exchanged onto the zeolite and also incorporated into the amorphous material.

Diffuse rqfectance spectroscopy. UV-Vis diffuse reflectance spectra were recorded on the three iron-exchanged zeolites. A peak at 240-250nm, as well as a band at 220 nm, was observed for all three zeolites. The peak at 290 nm is present in erionite and zeolite Y but does not appear on the mordenite spectrum. Mordenite exhibits a small absorption band 320 nm, which appears also in the erionite spectrum but is nonexistent in the zeolite Y spectrum. Finally, mordenite also exhibits a small absorbance at 450 nm.

The common peak at 240-250 nm may be assigned to the charge transfer transition between the oxygen atoms and the framework iron species, which explains why it is almost nonexistent in the case of zeolite Y, which does not have any framework iron. Tuel et al. assigned the 220 nm band in iron-containing mesoporous silicas to a ligand-to-metal charge transfer that involves isolated, tetrahedrally coordinated [Fe^{3+}]. The peak at 295 nm is of particular interest because it is present in the case of erionite and zeolite Y but does not appear in the mordenite spectrum. Conversely, mordenite exhibits a small absorption band at 320 nm, which also appears in the erionite spectrum but is nonexistent in the zeolite Y spectrum. This band region has been correlated with spin-forbidden d-d transitions in Fe-silicalite. Finally, the small absorbance of mordenite at 450 nm may be indicative of impurity iron oxides.

Fenton chemistry: The redox state of iron is relevant for Fenton-type reactions because iron(II) is necessary to initiate this reaction. Although the Haber-Weiss cycle allows the reduction of iron(III) by reducing agents (e.g., superoxides), we ensured that our sample began with iron in the +2 state. Therefore, we started with ferrous sulfate salts and used caution to keep iron in this redox state, following procedures given by previous researchers.

We used two methods to analyse for the hydroxyl radicals. The difference in the amount of DMSO before and after reaction with hydroxyl radicals was estimated from liquid chromatography. The second method involved UV-visible spectroscopy for estimating MSA formed by reaction of hydroxyl radicals with DMSO, and a detection limit of 100 lamol of MSA was achieved (24,25).

For mordenite and erionite, the production of hydroxyl radicals increases as surface iron increases. Increasing amounts of surface iron on zeolite Y did not give rise to increasing production of hydroxyl radicals. Results obtained with fractionated mordenite samples are similar to those with unfractionated mordenite for similar iron loading levels. For similar amounts of surface iron, and especially at the higher surface iron levels, erionite induced a larger hydroxyl radical production than mordenite, which induced a larger production than zeolite Y. The data obtained with free iron in solution show a relatively small production of hydroxyl radicals (7.3 x [10^{-5}] mmol for 350 [micro]g iron).

Measurement of the Oxidative Burst

The difference in results pertaining to the measurement of the oxidative burst between the chemiluminescence and the flow cytometry techniques can be rationalised as follows. Chemiluminescence technique provides a measure of [H_2][O_2], O[H^*], [$O_2^{*\cdot}$], whereas the flow cytometry, though supposedly specific to [H_2][O_2], has been shown recently to be sensitive to a variety of reactive oxygen species, including superoxides (36,37). In addition, it is known that the rate of spontaneous [$O_2^{*\cdot}$] dismutation into [H_2][O_2] is rapid, either spontaneously or by the enzyme superoxide dismutase. Therefore, the measurement of different oxygen species is not the cause for the differences in results. An examination of the reported detection limits for the chemiluminescence and flow cytometry method provide a better explanation. For the NR8383 cell line, Helmke et al. reported 4 x [10^{-9}] M [H_2][O_2]/5 x [10^5] cells after exposure to zymosan, which corresponds approximately to [10^{-17}] mol of [H_2][O_2] per cell. Since 100,000 cells are present in the chemiluminescence system, this would correspond to [10^{-12}] mol of [H_2][O_2] generated, or a [10^{-6}] M concentration (well volume is 250 [micro]L). The signal from zymosan is 10-100 times higher than the mineral-induced signals, which would set the concentration of hydrogen peroxide generated upon mineral stimulation in the [10^{-8}] to [10^{-7}] M range. Detection limits have been reported for chemilumene scence in the [10^{-8}] M range of concentrations for [H_2][O_2] , which is consistent with the detection of responses from mineral-challenged cells. In flow cytometry, the signal detected per cell has been reported to be [10^{-17}] mol of oxidised fluorescein dye per cell and is in the range generated upon zymosan stimulation. However, the [H_2][O_2] generated by exposure to minerals is several orders magnitude lower than that induced by zymosan and therefore below the detection limit of flow cytometry.

Phagocytosis and the Oxidative Burst

The three issues that we have discussed are cell viability upon exposure to the zeolites, the origin of the oxidative burst, and the magnitude of the oxidative burst for the three samples. A calculation of the number of particles in 250 [micro]g of zeolite Y, based on a density of 2.3 g/[cm^3] and a radius of 0.9 [micro]m (assuming a spherical shape) gives rises to an estimated value of 300 particles/cell. Although this number is high, cell viability is retained: Light microscopy showed similar numbers of living cells before and after exposure to particles. Furthermore, previous researchers have shown that a decrease in cell viability could be detected by a rapidly decreasing chemiluminescence signal, returning to baseline signals faster than in healthier cell systems. Such a phenomenon would be more likely to happen at higher fibre dosage. We have not observed a more rapidly decreasing signal at 250 [micro]g dosages when compared to the 10 [micro]g dosage. Therefore, cell viability is estimated to be similar for all three particles types and dosages in this time range.

We also have addressed the issue of the signal appearing from phagocytosis versus just membrane binding. For all three zeolites, the chemiluminescence peak typically occurred between 20 and 25 min and corresponds to the maximum in the oxidative burst. Nadeau et al. observed that the time at which the maximum chemiluminescence signal is obtained is inversely related to dosage. Urano et al. claimed that the earlier peaks could be reflecting faster kinetics in the superoxide generating process due to erionite fibre-membrane interaction rather than phagocytosis. In the present study, the time at which the peak maximum occurred did not vary after treatment of the cells by $N[H_4]Cl$. The decrease in the side scatter after treatment of the cells by the endocytosis inhibitor shows that phagocytosis is decreasing. This is consistent with the 70% decrease in chemiluminescence after treatment of the cells by ammonium chloride, suggesting that phagocytosis, rather than membrane interaction, is triggering the oxidative burst. These results correlate with the findings of Ng et al.

After addition of endocytosis inhibitors (chloroquine and ammonium chloride) to microsphere-exposed NR8383 cells, they observed no cell-microsphere interaction. They concluded that phagocytosis rather than simple binding caused the chemiluminescence signal. This phagocytosis-induced response of NR8383 cells, rather than a membrane response, is in accordance with the relatively small burst that Helmke et al. obtained with phorbol myristate acetate (PMA) stimulation compared to zymosan-induced stimulation of NR8383 cells. They found a maximum 2-fold increase in the chemiluminescence signal after PMA stimulation, and an average 23-fold increase after zymosan stimulation. Because PMA-induced response is a membrane phenomenon, while zymosan activation involves phagocytosis, this suggests that NR8383 cells do not show a large oxidative burst upon membrane stimulation. We have observed that all mineral-induced responses were 10- to 100-fold smaller than the zymosan response, which is consistent with previous experiments. Ng et al. reported small, dose-dependent signals of chemiluminescence upon uptake of polymeric microspheres by NR8383 cells, compared to the zymosan-induced chemiluminescence.

For the same mass loading of particulates, the signal from oxidative burst upon phagocytosis is inversely correlated with particle size, regardless of the nature of the zeolite. Fractionated mordenite induces a larger oxidative response than unfractionated mordenite. Similarly, the third size fraction of erionite, "fine erionite," induced significantly larger response than fractionated erionite, whereas fractionated erionite induced a larger response than ground erionite, all at a dosage of 50 [micro]g. Unfortunately, we do not know the number of particles in the cell upon exposure to samples with different particle sizes. If we assume that the same mass of particles is entering the cell, then there will be larger number of smaller particles per cell for the finer size and that would thus present a much larger surface area. In this case, it would imply that there is a positive correlation between the oxidative burst and the surface area of particles. As far as the differences in the magnitude of the oxidative burst for the different particles, it suggests that for comparable sizes of the particles, there is no significant difference in the oxidative burst, especially between the minerals erionite and mordenite.

Fenton Chemistry at the Mineral Surface: Methodology

Both methods used to estimate hydroxyl radicals involve the reaction of O[H^*] with DMSO followed by either measurement of the remaining DMSO or the amount of the stable product, MSA. The reaction of hydroxyl radicals with DMSO has an appreciable rate constant (k = 7 x [10^9] L/mol*sec), and the reported yield for this reaction varies from 85% to 91% (24,25,43). The MSA method gave somewhat lower values than the HPLC method, but the trends in both cases were similar. We believe that the MSA method is more accurate because it determines the amount of a product formed rather than measuring differences between the amount of the reactant. To confirm that the Fenton chemistry is solely due to surface-bound zeolitic iron rather than the bulk, measurement of the MSA production was done on iron-zeolite that was treated with Dab-4 polymer. On these zeolites, surface iron is absent, and iron present only in the bulk. No MSA production was observed.

The concern that $[H_2][O_2]$ and/or DMSO could diffuse into the zeolite cages and thereby exhibit a smaller signal in the HPLC method was examined by quantitatively estimating the levels of $[H_2][O_2]$ and DMSO after exposure to iron-containing zeolites. In all three zeolites, there was no decrease in amount of $[H_2][O_2]$. Erionite and mordenite showed no loss of DMSO, but zeolite Y did. This finding correlates well with dimension considerations: The kinetic diametre of DMSO is 5 [Angstrom], so it is size-excluded from erionite (pore windows 3.6 x 5.1 [Angstrom). DMSO can penetrate mordenite through the 6.5 x 7.0 [Angstrom] windows, but the defect structure limits the porosity. Zeolite Y showed a decrease of 1.09 x $[10^{-4}]$ [+ or -] 3 x $[10^{-6}]$ mmol of DMSO/g zeolite, and the data for the HPLC method were corrected for the pore trapping of DMSO. After correction, results comparable to the MSA method were observed.

Another concern was that iron released into solution from the zeolite was responsible for the Fenton chemistry. The released iron in solution after reaction with the DMSO/$[H_2][O_2]$ solution was estimated spectrophotometrically (phenanthroline method). This method is less sensitive than ICP-OES, but this latter technique was not an option because colloidal zeolite could be present in solution. Approximately 5% of the surface iron was found to be in solution. Since the aqueous complexes of iron show a considerably smaller "Fenton activity" than zeolite-bound iron, we rule out the possibility that solution iron plays an important part in the radical production.

Other complications could involve reaction of methyl radicals with DMSO, leading to overestimation of the results obtained by the HPLC method. This reaction will not influence the MSA method. The rate constant between $[H_2][O_2]$ and hydroxyl radicals, k = 1.2 - 4.5 x $[10^7]$ L/mol*sec, is smaller than the rate constant of $O[H^*]$ with DMSO (k = 7 x $[10^9]$ L/mol*sec) , thus, loss of hydroxyl radicals by reaction.with $[H_2][O_2]$ is not important. Ulanski et al. studied the kinetics of reaction of methyl radicals with $[H_2][O_2]$ and found a reaction rate of 2.7 x $[10^4]$ L/mol*sec, which is slow enough not to be problematic.

Role of Zeolite Surface Structure in Fenton Chemistry

For mordenite and erionite, the production of hydroxyl radicals increases as surface iron increases. Experiments with fractionated mordenite show that although surface iron increases per gram of zeolite, its activity in the Fenton reaction normalised to surface iron levels is similar to unfractionated mordenite. For comparable amounts of surface iron, especially at the higher loadings, erionite induces a larger hydroxyl radical production than mordenite, which induces a larger production than zeolite Y. Though the mechanism of the Fenton reaction is still being debated (47,48), it is accepted that the ligand around the iron influences its participation in the Fenton reaction. The ways in which a ligand can influence Fenton reactivity is by limiting access to $[H_2][O_2]$ as well as modification of the iron redox potential. We propose that the differences observed among the three zeolites are because varying iron coordination on the zeolite surface modifies its participation in the Fenton reaction.

Complexes with smaller reduction potentials such as EDTA (0.12 V) have higher rate constants (10 x $[10^3]$ L/mol*sec) for the Fenton reaction than complexes with higher reduction potentials such as aqueous complexes of Fe(II), 76 L/mol*sec (0.77 V) (47,48). However, Fe-EDTA has a higher rate constant (factor of 10) than Fe-diethylenetriamine pentaacetic acid (DTPA) (E° value of 0.03 V) and has been suggested to arise from unavailable coordination sites in DTPA. Even for thermodynamically unfavourable reactions, such as with complexes that show a redox potential higher than the reduction potential of the $[H_2][O_2]/O[H^-]$, $O[H^*]$ couple (0.32 V), such as N,N'-di-2-picolyl-4,7-diaza-l-oxacyclonanne (DPC), N,N',N"-tri-2-picolyl-l ,4,7-triazacyclononane (TPC), or water, with reduction potentials of 0.74, 0.71, and 0.77, respectively, Fenton activity is observed and the involvement of peroxo- complexes in an inner sphere mechanism has been proposed. This affinity would be higher for DPC than for hexa-aquocomplexes, which would explain the rate constants of 0.65 x $[10^3]$ and 0.076 x $[10^3]$ L/mol*sec, respectively, for H_2O and DPC. The differences in rates observed between Fe-DPC and Fe-TPC were attributed to the presence or absence of a free coordination site on the metal; DPC is a pentadentate ligand, TPC is a hexadentate ligand. Singh and Hider studied Fe(III) complexes and conduded that a high affinity of a ligand for Fe(III) would stabilise iron in this redox state, hence preventing it from participating further in the Fenton reaction. This is particularly clear in the case of the desferrioxamine ligand (DFO), for which the stability constant with Fe(III) is among the highqst (log K= 30.7) and for which the rate of substrate hydroxylation via Haber-Weiss mechanism is among the lowest.

Literature regarding the Fenton activity of mineral-bound iron is scarce. Studies have been performed on iron-exchanged mordenite and goethite. Joshi and Limye studied the catalytic decomposition of

$[H_2][O_2]$ by Fe(III)-mordenite, but did not conclude if the rate-determining step was the reduction of Fe(III) by the peroxide anion or the reduction of $[H_2][O_2]$ by Fe(II) (Fenton reaction). Reports of the reduction potential of iron in natural amphibole, magnetite and biotite range from 0.27 to 0.52 V, values subsequently lower than the reduction potential of iron in water (0.77 V).

The fact that a metal on different zeolites can exhibit different reduction potentials is also suggested by zeolite electrochemistry. The reduction potential of Cu(II)-exchanged zeolites using cyclic voltammetry show a 30 mV difference in half-wave potential ($[E_{1/2}]$) between the cyclic voltammograms of copper-exchanged mordenite and copper-exchanged zeolite Y. Because zeolite electrochemistry is a surface phenomenon, these results suggest a difference of the reduction potential for Cumordenite and Cu-zeolite Y.

Comparison of the differences in the structure of the zeolites is of interest. The a and b axes of erionite represent the fibre surface and shows a network of 8-membered rings (3.6 x 5.1 [Angstrom]). The natural mordenite sample is not primarily fibrous. Its surface shows channel openings made of 12-membered rings (7.0 x 6.5 [Angstrom]), 8-membered rings (2.6 x 5.7 [Angstrom]), and smaller 4-membered rings.

Zeolite Y has a more spherical morphology, and its surface presents 12-membered rings (7.4 [Angstrom] x 7.4 [Angstrom]). Iron on the zeolite surface is coordinated to the different aluminosilicate rings. The small amounts of hydroxyl radicals produced relative to the surface iron for all samples show that not all iron species on a zeolite surface are Fenton active.

This is supported by results from Fubini and co-workers (55-57), who have argued that only a few surface iron species on mineral samples are in the right redox and coordination state to be active in the hydroxyl radical generation. This correlates well with the various coordination environments (and their respective abundances) that a zeolite provides for binding iron. The increase in the amounts of radicals per surface iron for erionite is steeper than for mordenite, indicating that erionite surface sites provide a Fenton-enhancing coordination for iron. Zeolite Y shows little or insignificant activity and could arise from binding to the 12-membered rings, which would leave most of the iron coordination sites with water, and therefore its response is comparable to water-coordinated iron in generation of hydroxyl radicals.

Correlation of Biological and Chemical Reactivities

We reached two important conclusions after comparing the results of this study to those in the literature. First, for similar masses of particle exposure, the particle size (presumably by virtue of surface area) has a more significant effect on the oxidative burst upon phagocytosis than the mineral structure. Second, the Fenton chemistry that determines the production of hydroxyl radicals via iron on the zeolites depends on the mineral structure. Thus, while we found that erionite, mordenite, and zeolite Y have comparable oxidative bursts, hydroxyl radical production follows the order erionite > mordenite [much greater than] zeolite Y.

Although there is scientific consensus about the adverse effects of erionite in DNA strand breaks or mesothelioma induction, the cytotoxicity of mordenite and zeolite Y has not been extensively studied. Based on aerodynamic diametre calculations, Stephenson et al. estimated pulmonary deposition of fibre shaped-mordenite particles and concluded that substantial deep-lung deposition was possible. Hansen and Mossman compared in vitro superoxide generation induced by asbestos as well as erionite and mordenite and concluded that mordenite was less active than erionite. They attributed this difference to the fibrous shape of erionite. However, the size distribution of their samples was broad and different for the two zeolites, which casts some uncertainty on the correlation of structure with the oxidative burst. Adamis et al. studied in vitro and in vivo cytotoxicity of well-characterised mordenite samples. In vivo studies after intracheal instillation showed acute and subacute inflammation of the lung, which was attributed to the needle-shaped particles (10-12% of the mordenite sample)? In vitro studies showed significant hemolysis but moderate macrophage cell-membrane damage, as compared to quartz. However, the mordenite samples used in these studies could have contained impurities at levels of 45%.

The chemistry of the Fenton reaction at zeolite surface was investigated by Fubini and co-workers (55-57). They studied iron-exchanged zeolite Y as well as asbestos and iron oxides and concluded that-the extent of hydroxyl radical release did not parallel the mineral iron content or its surface iron content. They used the total iron content and specific surface area, which does not take into account the differences of iron exchange on different coordination sites. In the present study we show, for example, that although mordenite exchanges less total iron than erionite, the surface iron levels are similar. We agree, however, with Fubini et al.'s assessment that only a fraction of the iron species on the mineral surface is in the right redox state and coordination environment to participate in the Fenton reaction. (55-57). Thus, the considerable toxicity of erionite may be due to its fibrous nature and size, which ensures penetration into the lungs, and its surface chemistry, which promotes the formation of hydroxyl radicals. Mordenite, with comparable morphology, should also be toxic because of its efficacy in the Fenton reaction, whereas zeolite Y should be considerably less toxic because of its poor Fenton chemistry.

This study focused on two aspects of aluminosilicate mineral chemistry: interactions with macrophage cells (NR8383) and the ability of the mineral surface to produce hydroxyl radicals from $[H_2][O_2]$ (Fenton reaction). Using ammonium chloride as an endocytosis inhibitor, we have concluded that the dose-dependent chemiluminescence signal due to oxidative burst is mainly from phagocytised particles rather than to particles attached to the membrane. Using size-fractionated samples of the same mineral, we established that the generation of reactive oxygen species is inversely related to the size of the particles for similar levels (by mass) of particle exposure. Iron on the zeolite surface was estimated by exchange with a charged polymer. Hydroxyl radicals generated from $[H_2][O_2]$ via Fenton reaction and mediated by the Fe(II) zeolites were measured using a chromatographic and a spectrophotometric method. Iron on the surfaces of three different zeolites produced different amounts of hydroxyl radicals and followed the order erionite > mordenite > zeolite Y. We attribute the differences observed between the three zeolites as arising from different coordination of the iron on the zeolite surface. The biological implication of this study is that the size (surface area) of the mineral rather than its structure is the determining factor for the oxidative burst upon phagocytosis, whereas the structure of the mineral plays a key role in determining the production of iron-mediated hydroxyl radicals via the Fenton reaction.

Chapter 3

Cross Cellular Membranes by Nonphagocytic Mechanisms

High concentrations of airborne particles have been associated with increased pulmonary and cardiovascular mortality, with indications of a specific toxicologic role for ultrafine particles (UFPs; particles < 0.1 [micro]). Within hours after the respiratory system is exposed to UFPs, the UFPs may appear in many compartments of the body, including the liver, heart, and nervous system. To date, the mechanisms by which UFPs penetrate boundary membranes and the distribution of UFPs within tissue compartments of their primary and secondary target organs are largely unknown. We combined different experimental approaches to study the distribution of UFPs in lungs and their uptake by cells. In the in vivo experiments, rats inhaled an ultrafine titanium dioxide aerosol of 22 nm count median diametre. The intrapulmonary distribution of particles was analysed 1 hr or 24 hr after the end of exposure, using energy-filtering transmission electron microscopy for elemental microanalysis of individual particles. In an in vitro study, we exposed pulmonary macrophages and red blood cells to fluorescent polystyrene microspheres (1, 0.2, and 0.078 [micro]m) and assessed particle uptake by confocal laser scanning microscopy. Inhaled ultrafine titanium dioxide particles were found on the luminal side of airways and alveoli, in all major lung tissue compartments and cells, and within capillaries. Particle uptake in vitro into cells did not occur by any of the expected endocytic processes, but rather by diffusion or adhesive interactions. Particles within cells are not membrane bound and hence have direct access to intracellular proteins, organelles, and DNA, which may greatly enhance their toxic potential. Key words: aerosol, erythrocytes, lungs, macrophages, microscopy, nanoparticles, rats, surfactant.

High concentrations of airborne particles have been associated with increased pulmonary and cardiovascular mortality, with indications of a specific toxicologic role for ultrafine particles (UFPs; particles with diametres < 0.1 [micro]m). UFPs may induce inflammatory and prothrombotic responses, promoting atherosclerosis, thrombogenesis, and the occurrence of other cardiovascular events. Human data suggest that inhaled UFPs influence lung physiology. UPFs may also affect the autonomic nervous system or act directly on cells in various organs and induce mutations. After exposure of the respiratory system to UFPs, the UFPs may appear within hours in many compartments of the body, including the liver, heart, and nervous system.

UFPs are formed by gas-to-particle conversion or by incomplete fuel combustion. Despite considerable efforts to reduce air pollution, the environmental burden by UFPs may have increased rather than decreased over time. Moreover, the fast-growing nanotechnology industry generates new UFPs daily, which may become aerosolised at some stage and may present additional health risks. UFPs possess increased toxicity compared with larger particles composed of the same materials. Their environmental burden is characterised by high number concentrations but low mass concentrations. Thus, a relatively large surface area per unit mass facilitates adsorption of various organic compounds from the ambient air and enhances interaction with biological molecules within the organism.

Deposition of UFPs in the respiratory system is caused by diffusional displacement. Depending on particle size, deposition occurs efficiently in the nose, the conducting airways, and the alveoli. Although particles with diametres > l [micro]m usually remain on the epithelial surface upon their deposition and are subjected to clearance by cough, mucociliary transport, and/or phagocytosis by macrophages, UFPs seem to penetrate the boundary membranes of the lungs rapidly a unique feature for insoluble particles. In addition, transport across the olfactory epithelium and accumulations in the brain were reported for various UFP types (Oberdorster 2004). In vitro experiments revealed penetration of UFPs into mitochondria of macrophages and epithelial cells that was associated with oxidative stress and mitochondrial damage.

Because everyone on earth inevitably inhales thousands to millions of UFPs with each breath, it is important to assess health risks by UFP air pollution. The costs of actions to be taken to reduce ambient aerosol particles are high and will affect the economy greatly, presenting an urgent need to clarify the fate of inhaled UFPs. To date, the mechanisms by which UFPs penetrate boundary membranes and the distribution of UFPs within tissue compartments of their primary and secondary target organs are largely unknown.

This study is the first to investigate the distribution of inhaled UFPs within lungs at the individual particle level and combines different experimental approaches an in vivo inhalation study in rats and an in vitro cell exposure study on pulmonary macrophages and red blood cells (RBCs).

In the in vivo experiments, rats inhaled an ultrafine titanium dioxide aerosol of 22 nm count median diametre (CMD) during 1 hr, resulting in a deposition of 4-5 [micro]g Ti[O_2] per animal. The intrapulmonary distribution of deposited particles was analysed immediately or 24 hr after the end of exposure, using energy-filtering transmission electron microscopy (EFTEM) to allow elemental microanalysis of individual particles. In the in vitro study, we exposed cultured porcine pulmonary macrophages and human RBCs to fluorescent polystyrene microspheres with diametres of 1, 0.2, and 0.078 [micro]m and assessed particle uptake by confocal laser scanning microscopy (CLSM).

The animal experiments were conducted under federal guidelines for the use and care of laboratory animals (German Animal Protection Law) and were approved by the District of Upper Bavaria (Approval No. 211-2531-108/99) and by the GSF Institutional Animal Care and Use Committee, as well as in accordance with the Swiss Federal Act on Animal Protection and the Swiss Animal Protection Ordinance. Ten young, adult, male WKY/NCrl BR rats [body weight (bw) 250 [+ or -] 10 g (mean [+ or -] SD; Charles River, Sulzfeld, Germany] were housed under standard conditions, with access to food and water ad libitum, in a room controlled for humidity (55% relative humidity) and temperature (22°C) and with lighting on a 12-hr day/night cycle. Animals were anesthetized by intramuscular injection of a mixture of medetomidine (Domitor; 150 [micro]g/100 g bw; Pfizer GmbH, Karlsruhe, Germany), midazolam (Dormicum; 0.2 mg/100 g bw; Hoffmann-La Roche AG, Grenzach-Wyhlen, Germany), and Fentanyl (0.5 [micro]g/100 g bw; Janssen-Cilag GmbH, Neuss, Germany) for inhalation and lung fixation and anticoagulated by intraperitoneal injection of 2,000 IU heparin (Heparin-Natrium-ratiopharm, ratiopharm GmbH, Ulm/Donautal, Germany) for lung fixation. Anesthesia was antagonised by subcutaneous injection of atipamezole (Antisedan; Pfizer GmbH), flumazenil (Anexate, Hoffmann-La Roche

AG), and naloxone (Narcanti, Janssen Animal Health, Neuss, Germany).

Aerosols and inhalation: Titanium is suitable for EFTEM because it does not interfere with the heavy metals used for tissue preparation. $Ti[O_2]$ is inert and nonpathogenic, and ultrafine aerosols thereof remain as insoluble particles. Generally, commercially available $Ti[O_2]$ particles show a positive Zeta potential of 20-30 mV as measured by Malvern Zetasizer 3000 HS. The generation and inhalation of the $Ti[O_2]$ aerosol used in this study has been described in detail by Kapp et al. (2004). Briefly, ultrafine $Ti[O_2]$ aerosols were generated with a Palas spark generator in a pure argon plus 0.1% oxygen stream. The aerosol was quasi-neutralised by a radioactive ^{85}Kr source, diluted, and conditioned for inhalation. Particle size distribution and number concentration were monitored continuously by a differential electrical mobility particle sizer and a condensation particle counter. An aerosol of 22 nm CMD (geometric SD of 1.7) was produced. The mean number concentration was 7.3 x $[10^6]$ particles/$[cm^3]$ (SD 0.5 x $[10^6]$ particles/$[cm^3]$), corresponding to a mass concentration of 0.11 mg/$[m^3]$. The 22 nm particles measured as aerosol particles are already agglomerates of smaller primary structures formed immediately after spark ignition and condensation. The estimated size of the primary structures was 4 nm as derived from the measured specific surface area of 330 $[m^2]$/g of the UFPs produced in this study and the $Ti[O_2]$ bulk density of 4.2 g/$[cm^3]$.

Each anesthetized rat was placed in an airtight plethysmograph box. The animals inhaled the aerosol in pairs (one each for the 1-hr and 24-hr examinations) for 1 hr via an endotracheal tube by negative-pressure ventilation (-1.5 Pa) at a breathing frequency of 45/min, resulting in a minute volume of about 200 $[cm^3]$/min.

From breathing and aerosol parameters, the deposited amount of $Ti[O_2]$ was calculated to be 4-5 [micro]g in each rat. After aerosol exposure, anesthesia of one of the paired rats was antagonised as described above, and the animal was returned to its cage for 24 hr.

Lung fixation and tissue preparation: Lungs were fixed either 1 hr or 24 hr after the aerosol inhalation by sequential intravascular perfusion with buffered 2.5% glutaraldehyde (Agar Scientific Ltd., Piano GmbH, Wetzlar, Germany), 1.0% osmium tetroxide (Simec, Zofingen, Switzerland), and 0.5% uranyl acetate (Fluka Chemie GmbH, Sigma-Aldrich, Buchs, Switzerland). Lungs were then subjected to systematic tissue sampling, dehydration in a graded series of ethanol, and embedding in Epon (Fluka). Ultrathin ([less than or equal to] 50 nm) sections were cut from five to eight tissue blocks per animal, mounted onto uncoated 600-mesh copper grids, and stained with uranyl acetate and lead citrate (Ultrostain, Leica, Glattbrugg, Switzerland).

Particle localisation and elemental microanalysis in situ: On ultrathin sections, 12 fields, corresponding to an area of 1,820 [micro][m^2] each, were systematically subsampled and investigated for the presence and localisation of Ti[O_2] particles in a LEO 912 transmission electron microscope (LEO, Oberkochen, Germany) equipped with an in-column energy filter allowing energy dispersion for element specific contrast. Ti[O_2] particles were identified by parallel electron energy-loss spectroscopy (parallel-EELS), electron spectroscopic imaging, and image-EELS. For elemental microanalysis, we used the [$L_{2,3}$] edge of Ti at 456 eV energy loss. We obtained bright-field and structure-sensitive micrographs (recorded at 250 eV), as well as element-specific contrast for Ti[O_2], by digital acquisition.

Cell culture experiments: Porcine lung macrophages (provided by K. McCullough and H. Gerber, Institute for Virus Diseases and Immune Prophylaxis, Mittelhausern, Switzerland) were cultured at [10^6] cells/mL in two-chamber slides (VWR International AG, Dietikon, Switzerland) for 24 hr at 37°C and 5% CO_2 in RPMI 1640 medium (containing 25 mM Hepes; LabForce AG, Nunningen, Switzerland) with 10% fetal bovine serum (LabForce), 1% L-glutamine (LabForce), and 1% penicillin/streptomycin (Gibco BRL, Invitrogen AG, Basel, Switzerland). Human RBCs, always freshly isolated from the same donor, were cultured at 8 x [10^6] cells/mL and for 6-24 hr, as described above.

For CLSM, fluorescent polystyrene microspheres with diametres of 1, 0.2, and 0.078 [micro]m (Fluoresbrite plain yellow green; Polysciences, Chemie Brunschwig AG, Basel, Switzerland) were added to the cells at [10^{10}] particles/mL in supplement-free RPMI 1640 medium for 4 hr. To inhibit phagocytic uptake, cells were pretreated with cytochalasin D (cytD, 10 [micro]g/mL; Fluka) for 30 min and during incubation with particles. Macrophages were then washed in phosphate-buffered saline (PBS), fixed in 3% paraformaldehyde/PBS, treated with 0.1 M glycine/PBS, permeabilised with 0.2% Triton X-100/PBS, and stained for F-actin with rhodamine-phalloidin (1:100; Molecular Probes, VWR International AG, Lucerne, Switzerland) for 60 min at 37°C. RBCs were fixed in 2.5% glutaraldehyde/PBS, resulting in a red autofluorescence. Preparations were mounted in PBS:glycerol (2:1) containing 170 mg/mL Mowiol 4-88 (Calbiochem, VWR International AG).

For TEM analysis, RBCs were incubated with 0.025 [micro]m gold particles at 6.6 x [10^{10}] particles/mL. RBCs were fixed with buffered 2.5% glutaraldehyde, 1% osmium tetroxide, and 0.5% uranyl acetate, and prepared for TEM as described above for lung tissue.

Microscopic analysis of cultured cells: We used a MicroRadiance system from Bio-Rad (Hemel Hempstead, UK) combined with an inverted Nikon microscope. Optical sections with a voxel dimension of 50 x 50 x 200 nm were taken with a 100x/1.4 plan-apochromate objective. We performed image processing and visualisation using IMARIS software. We applied a deconvolution algorithm using Huygens software to increase axial and lateral resolutions and to decrease noise. Experiments were performed in triplicate or quadruplicate, and 30-50 cells were scanned by CLSM for each data point. We examined ultrathin sections of RBCs incubated with gold particles in a Philips 300 TEM at 60 kV.

Distribution of inhaled UFPs in lungs: We found Ti[O_2] particles on the luminal side of airways and alveoli as well as within each tissue compartment of the lung. Particles were localised within epithelial and endothelial cells, within fibroblasts and between collagen fibrils in the connective tissue, within blood capillaries, and even within RBCs. Intracellular particles were localised most often in the cytoplasm and rarely within the nucleus. Particles within cells were not membrane bound. On average, 79.3 [+ or -] 7.6% (mean [+ or -] SD) of the particles were found on the luminal side of airways and alveoli, 4.6 [+ or -] 2.5% were within epithelial or endothelial cells, 4.8 [+ or -] 4.5% within the connective tissue, and 11.3 [+ or -] 3.9% within the capillaries. The relative distributions of particles among different lung compartments at 1 hr and at 24 hr after inhalation were not different from each other and correlated with the volume fractions of the respective compartments.

Uptake of UFPs by macrophages and RBCs: We studied the uptake of fine (1-0.2 [micro]m) and ultrafine (< 0.1 [micro]m) fluorescent microspheres in cultured macrophages as a model for phagocytic cells and in human RBCs as a model for nonphagocytic cells. Additionally, we treated macrophage cultures with cytD to block phagocytosis. Cells were prepared for CLSM, and a deconvolution algorithm was applied to increase lateral as well as axial resolution. We found particles of all sizes within macrophages, however, with different percentages of cells involved in particle uptake. On average, 77 [+ or -] 15% (mean [+ or -] SD) of the macrophages contained UFPs, 21 [+ or -] 11% contained 0.2 [micro]m particles, and 56 [+ or -] 30% contained 1 [micro]m particles. CytD did not inhibit the uptake of ultrafine and 0.2 [micro]m particles but blocked the uptake of 1 [micro]m particles by these cells. We found ultrafine and 0.2 [micro]m, but not 1 [micro]m, particles in RBCs. TEM analysis of RBCs incubated with 0.025 [micro]m gold particles showed that intracellular particles were not membrane bound.

The ultrastructural analyses of lung tissue demonstrated that I hr after aerosol inhalation, 24% of ultrafine TiO_2 particles, on average, were located within and beyond the epithelial barrier (i.e., in the main lung tissue compartments, in the cytoplasm, and in the nucleus of the cells). These results confirm data from human studies, where the inhalation of ultrafine carbon particles affected pulmonary diffusing capacity, suggesting that particles in the interstitium have physiologic effects.

This study also provides evidence for some particle translocation into the microvasculature. In previous studies on iridium particles, we found minute fractions of particles translocated into secondary target organs. Different particle materials iridium versus TiO_2 may have resulted in different particle translocation patterns, and we do not know the exact composition and structure of the ultrafine particle surface, which is likely to influence particle translocation. The present study focuses on particle distribution within the primary target organ, the lung, and the results do not determine which fraction of particles may have escaped the lung microvasculature to be systemically circulated.

Particles found within cells were not membrane bound, indicating a nonendocytic uptake. In addition, the overall distribution pattern of the particles in the lungs (i.e., the percentages of particles in the different lung compartments) at 1 hr and at 24 hr after particle inhalation was the same and was correlated with the volume densities of the corresponding lung compartments, implying that ultrafine TiO_2 particles can move between tissue compartments without restraint. Our results are in contrast with those obtained by Stearns et al. (2001), who studied the uptake of ultrafine TiO_2 particles in vitro in the A549 epithelial cell line. They found that membrane-bound vesicles contained mostly large aggregates of TiO_2. Sometimes vesicles with clusters consisting of as few as two to three particle profiles were observed. In a pilot study in vitro with porcine macrophages, we obtained similar results.

Whether these clusters contained few particles only or were sectioned through the top of larger ones is not known. However, because ultrafine TiO_2 aggregate very quickly in polar liquids such as cell culture medium and because particle concentration was fairly high, as seen from micrographs, it is very likely that particles aggregated within the cell culture medium and that cells engulfed these large clusters by an endocytic pathway. Particle agglomeration on the lung epithelium during the 1-hr inhalation, however, is very unlikely because of the size of the inner lung surface and the number of deposited UFPs.

The fact that 80% of the retained TiO_2 particles were still on the luminal side of the epithelium even 24 hr after inhalation is surprising and in contrast to previous studies on the lavageability of ultrafine iridium particles, where only 20% of the particles could be lavaged from the epithelial surfaces 24 hr after inhalation. The low lavageability of ultrafine iridium particles in contrast to high lavageability of 80% of 0.5-10-[micro]m particles was interpreted as either higher adhesion of UFPs to epithelial membranes or epithelial uptake and penetration of UFPs into the interstitium.

The large fraction of TiO_2 particles on the luminal side of the epithelium in the present study strongly supports higher adhesion of UFPs to epithelial structures. However, it must also be considered that proteins are very likely to bind rapidly to the particles, which then may affect the further metabolic fate of the particles in terms of their adhesion, residence time on the epithelium or uptake, and even penetration through the epithelium. Along this line, the difference of the two particle materials and surfaces may have led to binding of those proteins, which then may have mediated major uptake and penetration of iridium particles into and through the epithelium. We already have first evidence that ultrafine commercial TiO_2 particles bind more readily to other proteins in the lung-lining fluid than carbonaceous and amorphous silica particles.

Microscopic analyses of phagocytic and nonphagocytic cells incubated with different particle types showed that macrophages take up fine and uhrafine polystyrene microspheres and that treatment with cytD inhibits the uptake of 1.0-[micro]m particles, but not uptake of the smaller particles, by these cells. Ultrafine polystyrene and gold particles also entered RBCs and were not membrane bound.

The mechanisms of intracellular uptake of macromolecules, particles, and even cells are subsumed as endocytosis. Material to be ingested is progressively enclosed by the plasma membrane, which eventually detaches to form an endocytic vesicle. Phagocytosis and pinocytosis are distinguished by the size of endocytic vesicles formed. Phagocytosis, a receptor-mediated, actin-based process, is characteristic for neutrophils, macrophages, and dendritic cells. It is the main mechanism for the clearance of insoluble 1- to 3-[micro]m particles from the alveoli. Pinocytosis involves the ingestion of fluid and solutes via vesicles of about 100 nm in diametre. There are at least four basic mechanisms, most of which can be demonstrated in lungs and involve specific receptor-ligand interactions: a) macropinocytosis, b) clathrin-mediated, actin-based endocytosis, c) caveolae-mediated endo- or transcytosis, and d) clathrin- and caveolae-independent endocytosis. Kruth et al. (1999) described an additional endocytic process, patocytosis, in which hydrophobic polystyrene particles < 0.5 [micro]m are transported through induced plasma membrane channels into an extensive labyrinth of interconnected membrane-bound compartments. None of these endocytic pathways, all of which include vesicle formation, is likely to account for the translocation of UFPs in our study, as intracellularly localised particles were not membrane bound. Moreover, because RBCs contained UFPs and cytD treatment of macrophages did not prevent UFP translocation into these cells, particle uptake by any actin-based mechanism can also be excluded.

Transport via pores, as suggested for lung-blood substance exchange, is another potential mechanism for UFP translocation. Ti[O_2] particles may diffuse through such pores. A transport mechanism by diffusion is consistent with the observed spatial distribution of UFPs in our inhalation study. Thus far, signal-mediated transport via pores has been demonstrated only for ultrafine gold particles of up to 39 nm in diametre, through the nuclear pore complex in Xenopus oocytes, where transport velocities depended on particle size.

Passive uptake (not triggered by receptor-ligand interactions) may also occur by electrostatic, Van der Waals, or steric interactions, subsumed under "adhesive interactions". Rimai et al. showed that 8-[micro]m glass particles were approximately 90% engulfed by a polystyrene substrate, compared with 22-pro particles, which were only 30% engulfed. However, the influence of particle size on their engulfment was not clarified in these publications. Several concepts for the nonspecific engulfment of particles through interfacial structures (including cell membranes) have been suggested. A thermodynamic model using the "wettability criterion" was successful in predicting passive particle uptake, although it did not take into account the elastic properties of the cell membrane. In another thermodynamic analysis combined with a molecular dynamics simulation, Bresme and Quirke (1999) showed that line tension influences the wetting bahaviour of nanoparticles at liquid-vapour and liquid-liquid interfaces. These authors found negative line tension values for particles of a few nanometres in diametre, but positive values for those an order of magnitude larger. A negative line tension favours the initial wetting of a spherical particle after its approach to an interface.

Shanahan (1990) studied the engulfment of solid-surface heterogeneities equivalent to particles in the nanometre range by unbalanced capillary forces (free energy perturbations). Thermal capillary waves cause fluid droplets to coalesce with a fluid substrate by film drainage at the interface, breakage of the film, and intrusion of the particle into the bulk phase. Thus, thermal capillary fluctuations may enhance particle transport through cell membranes. Experimental results demonstrate consistently greater immersion of smaller particles than larger ones into a liquid substrate covered by surfactant film in vitro as well as in situ in airways. These results support the concept that line tension plays a significant role in particle displacement.

It remains to be determined which chemical and physical properties of membranes and particles are responsible for the translocation of UFPs in vivo. Interestingly, we did not see any difference in particle uptake in vitro with respect to differing surface charges or surface chemistry when we used three different particle types: a metal, a metal oxide, and a synthetic polymeric material. However, these particles were added to the cells in suspension and did not approach the cells from the air or require passage through a surfactant film first, as in the in vivo inhalation experiments. In these in vivo experiments, electrostatic interactions are likely important for particle deposition and subsequent retention.

In summary, UFPs of various materials can cross any cellular membrane, but neither endocytosis, which is based on vesicle formation, nor any actin-based mechanisms are likely to account for UFP translocation into the cell. Our results from the inhalation experiments with $Ti[O_2]$ particles point to a transport mechanism that includes adhesive interactions or, in terms of thermodynamics, interfacial and line tension effects. In addition, particle diffusion and uptake promoted by thermal capillary waves might play a role in particle transport through membranes. After the deposition of nanometre-size particles, their further fate may be largely independent from particle surface chemistry and charge.

Consequently, it is a possible fate of inhaled ambient UFPs that they are transferred from the lungs to most other organs. In a first analysis of ultrathin sections from hearts of the same rats, we found ultrafine Ti[O_2] particles in the connective tissue, that is, within fibroblasts. There may be no means on the cellular level to prevent, influence, or direct their uptake. Moreover, the toxic potential of UFPs is greatly enhanced by their free location and movement within cells, which promote interactions with intracellular proteins and organelles and even the nuclear DNA.

Potential health implications of our findings are related not only to ambient UFPs but also to engineered "nanoscaled particles," which may be released into our environment during their production, transport, and aging, or during waste disposal. Routes of exposure to nanomaterials include oral, cutaneous, and inhalarive uptake, the latter being addressed in the present study. Although the number of particles translocated into the cells may vary substantially according to their physicochemical properties, the data of the present study strongly suggest that adverse health outcomes associated with the uncontrolled presence of nanoscale particles in tissues require further attention.

Implications in Epidemiologic Research

Epidemiologic research has shown increases in adverse cardiovascular and respiratory outcomes in relation to mass concentrations of particulate matter (PM) [less than or equal to] 2.5 or [less than or equal to] 10 [micro]m in diametre (P[$M_{2.5}$], P[M_{10}], respectively). In a companion, we discuss epidemiologic evidence pointing to underlying components linked to fossil fuel combustion. The causal components driving the PM associations remain to be identified, but emerging evidence on particle size and chemistry has led to some dues. There is sufficient reason to believe that ultrafine particles < 0.1 [micro]m (UFPs) are important because when compared with larger particles, they have order of magnitudes higher particle number concentration and surface area, and larger concentrations of adsorbed or condensed toxic air pollutants (oxidant gases, organic compounds, transition metals) per unit mass. This is supported by evidence of significantly higher in vitro redox activity by UFPs than by larger PM. Although epidemiologic research is needed, exposure assessment issues for UFPs are complex and need to be considered before undertaking investigations of UFP health effects. These issues include high spatial variability, indoor sources, variable infiltration of UFPs from a variety of outside sources, and meteorologic factors leading to high seasonal variability in concentration and composition, including volatility. To address these issues, investigators need to develop as well as validate the analytic technologies required to characterise the physical/chemical nature of UFPs in various environments. In the present review, we provide a detailed discussion of key characteristics of UFPs, their sources and formation mechanisms, and methodologic approaches to assessing population exposures.

Ambient aerosols are defined as suspensions of relatively stable solid or liquid particles in ambient air. Ambient particles range from a few nanometres to 100 lam in diametre. Ambient particulate matter (PM) has been described in three size distributions: ultrafine particles (UFPs), approximately < 0.1 [micro]m in diametre and largely consisting of primary combustion products; accumulation-mode particles between 0.1 and 2.5 [micro]m in diametre, from aggregation of UFPs and vapours; and coarse-mode particles > 2.5 [micro]m in diametre, largely mechanically generated particles.

Most ambient PM mass is distributed in the last two size ranges (or modes), that is, the coarse and accumulation modes. Typically, the mass-based size distribution of ambient PM is bimodal, with a "saddle" point in the size range of 1-3 [micro]m, which distinguishes the coarse and accumulation modes: by convention, the coarse mode consists of particles > 2.5 [micro]m in aerodynamic diametre, whereas the fine (or accumulation) mode consists of particles [less than or equal to] 2.5 [micro]m in aerodynamic diametre (P[$M_{2.5}$]). The particle size range < 0.1 [micro]m, known as the ultrafine mode, contains the majority (in numbers) of the ambient particles and an appreciable portion of total surface area.

These three particle modes have distinctly different chemical composition, sources, and lifetimes in the atmosphere. Particles in the coarse mode are produced by mechanical processes (grinding, erosion, and resuspension by the wind). Because of their relatively larger size, they have higher gravitational settling velocities and are thus removed from the atmosphere within hours. Particles in the accumulation mode are mostly anthropogenic in origin; they are generated through gas-to-particle conversion mechanisms, including homogeneous and heterogeneous nucleation, and by condensation onto preexisting particles in the accumulation-size mode. Because they are too small to settle out, particles of the accumulation mode have lifetimes in the atmosphere on the order of days, and they can be transported over long distances. The major chemical constituents of fine particles are sulfate, nitrate, ammonium, organic carbon, and elemental carbon (EC), as well as a variety of trace metals formed in combustion processes.

Because of their increased number and surface area as well as their high pulmonary deposition efficiency, UFPs are particularly important in atmospheric chemistry and environmental health. For example, the UFP's surface can carry large amounts of adsorbed or condensed toxic air pollutants (oxidant gases, organic compounds, and transition metals). Many of these toxic air pollutants have been identified as having pro-inflammatory effects, yet relevant exposure data are rarely available to epidemiologists. Results from several recent studies in mostly urban areas showed that a large proportion of urban UFPs consists of primary combustion products from mobile source emissions (particularly diesel and automobile exhaust) and includes organic compounds, EC, and metals. Because exposure to mobile emissions can vary across short distances and depends on personal activity patterns, assessing such exposures requires methods that go beyond the use of government monitoring data alone.

In this chapter we provide a detailed discussion of key characteristics of UFPs, their sources, and formation mechanisms and discuss methodologic approaches to the assessment of population exposures.

Characteristics and Potential Importance of UFPs

Unlike coarse and fine particles, which, as discussed above, are naturally divided by a cut-point of 2.5 [micro]m, there is no clear cut-point that separates UFPs from accumulation-mode PM. This is because, unlike coarse and fine (accumulation plus ultrafine) particles, which have distinctly different origins, a major fraction of accumulation-mode PM originates from the ultrafine mode. The distinction between the ultrafine and accumulation modes has varied from 0.1 to 0.2 [micro]m, depending on location and season for the following reasons.

"Ultrafine PM" is traditionally defined as particulates originating mostly from "fresh" emission sources and accounting for > 90% of the number-based particle concentrations. Recent studies in the Los Angeles Basin have shown that median mobility diametres in the inland valleys (downwind receptor areas) of the basin are in the 90-150 nm range in the summer months. This finding implies that a simple 100-nm cut-point does not always ensure that the vast majority of ambient particles by number concentration will fall below that size. The definition of UFPs is complicated further because unlike coarse- or accumulation-mode PM, for which the assumption that they have a perfectly spherical shape is reasonable, a significant fraction of directly emitted soot particles in the ultrafine range has a fractal or agglomerate-like structure. These particles are generated primarily from high-temperature combustion sources such as motor vehicles. By their very nature, agglomerate structures have higher surface areas than spherical particles with the same equivalent diametre and are generally less dense.

This deviation from a spherical shape creates a significant deviation between the aerodynamic diametre (typically measured by means of inertial classifiers, e.g., impactors) and mobility diametre (typically measured by means of electrical differential mobility analysers). Because of their low density, a substantial fraction of these particles would be classified by an inertial separator as UFPs, whereas an electrical mobility classifier would classify these irregular particles in larger size ranges because of their high surface area and, hence, mobility. These arguments have been further supported by observations in a recent study by McMurry et al. (2002) in which the effective density of diesel particles was measured by relating the mobility-measured diametre of combustion particles to their aerodynamic diametre. This study demonstrated that as the mobility size increases, the effective density tends to decrease, presumably because of the surface irregularities of the larger particles. It should be noted, however, that the relative abundance of these fractal-like agglomerates is highly variable, depending on the sampling location(s) as well as the time of day in order to account for the effect of vehicular emissions.

To date, there has been rapidly increasing epidemiologic evidence linking respiratory health effects and exposures to UFPs. Epidemiologic studies conducted by Peters et al. (1997) have demonstrated a stronger association between respiratory health in asthmatic adults and exposure to UFPs compared with fine or coarse particles. A study by

Pekkanen et al. (1997) showed associations of fine particles and UFPs with deficits in peak expiratory flow among asthmatic children. A study by Penttinen et al. (2001) also demonstrated a negative association between daily mean number concentrations (dominated by UFPs) and peak expiratory flow. Wichmann et al. (2000) showed that positive associations of cardiovascular mortality with both UFPs and fine ambient particles were comparable and seemed to be largely independent of each other in two-pollutant models. It has been proposed that UFPs have contributed to other epidemiologic findings of adverse effects of particulate air pollution on cardiovascular health. In a companion article, we discuss epidemiologic evidence indirectly pointing to the potential adverse cardiovascular effects of pollutant components of fossil fuel combustion that dominate the ultrafine fraction.

There is sufficient toxicologic basis for believing that UFPs are capable of inducing the greatest amount of inflammation per unit PM mass because of high particle number (PN), high lung deposition efficiency, and surface chemistry. UFPs have very low mass but magnitudes higher PNs and therefore a high surface area relative to fine or coarse particles. For example, a study in Pasadena, California, by Hughes et al. (1998) found that UFP PNs were consistently in the range of 1.3-8.9 x [10^4] particles/[cm^3] air, and UFP mass was only in the range of 0.80-1.58 [micro]g/[m^3]. Chalupa et al. (2004) found the deposition efficiency of carbon UFPs in human subjects was > 60% and increased with exercise and in subjects with asthma. As discussed below, toxic air pollutants carried by UFPs are expected to induce inflammatory responses through reactive oxygen species (ROSs) or other mechanisms.

Evidence now supports the view that UFPs carry considerable amounts of air toxics. Kim et al. (2002) studied composition of size-fractionated particulate air pollution in urban sites of Los Angeles, California, with considerable mobile source emissions. They found a large proportion of UFPs are made up of organic carbon, followed by EC as primary products from mobile source emissions, particularly diesel and automobile exhaust. Other studies showed that UFPs contain the largest fraction of polycyclic aromatic hydrocarbons (PAHs) by mass. Overall PAH concentrations are likely to be higher where there is a greater traffic density, such as downtown Los Angeles where diesel exhaust was found to make up 32.7% of the fine particle mass. Given these findings, if PAHs and other organic compounds are major causal components of the inflammatory response to PM, then greater responses from UFPs compared with larger particle fractions are expected at urban areas in the proximity of mobile sources. This expectation is further supported by the greater PN and surface area of UFPs, and greater internal doses due to the higher respiratory deposition of UFPs.

Experimental data show that compared with larger particles, UFPs are capable of avoiding phagocytosis by alveolar macrophages and gain entry to pulmonary interstitial sites, including vascular endothelium. Therefore, UFPs may induce pulmonary inflammation at both epithelial and interstitial sites, as well as enter the circulation to reach other target sites, including the cardiovascular system. Additionally, organic components of PM such as PAHs, which comprise a large proportion of both freshly emitted exhaust and secondary aerosols, have been shown to induce a broad polyclonal expression of cytokines and chemokines in respiratory epithelium. As discussed, this effect may be due to the action of metals, PAHs, and related compounds that lead to the production of cytotoxic ROSs. ROSs induce oxidant injury and inflammatory responses, including the production of nuclear transcription factor [kappa]B, which increases the transcription of cytokines and acute phase proteins.

These inflammatory and oxidant stress responses are expected to occur at extrapulmonary sites as well, including the vascular endothelium of the heart. Evidence for the importance of oxidant stress responses to cardiovascular effects is that antioxidant therapy is protective against the development of hypertension, aetherosclerosis, cardiomyopathies, coronary heart disease, and congestive heart failure. Li et al. (2003) showed that UFPs in Los Angeles were most potent toward inducing cellular heme oxygenase-1 (HO-1) expression and depleting intracellular glutathione, both important in oxidant stress responses. A separate study in Los Angeles by Cho et al. used the dithiothreitol assay as a quantitative measure of in vitro ROS formation. That study also showed that UFPs had the highest ROS activity. Li et al. (2003) also showed that UFPs and, to a lesser extent, accumulation-mode particles localise in mitochondria where they induce major structural damage, which may contribute to oxidative stress. Xia et al. (2004) showed that suspensions of urban UFPs as well as diesel exhaust particles on epithelial cells decreased membrane potential and induced both loss of mitochondrial membrane mass and apoptosis. Interestingly, commercial polystyrene nanoparticles failed to exert a mitochondrial effect. Together, these studies provide strong evidence that the increased biologic potency of UFPs is related to the content of redox cycling organic chemicals and ability to damage mitochondria.

Sources and Formation Mechanisms of UFPs

UFPs may be formed in the atmosphere by at least three processes: a) UFPs may be formed during combustion processes associated mostly with traffic or industrial sources, and emitted directly to the atmosphere as UFPs; b) these combustion processes may also emit hot supersaturated vapours, which undergo nucleation and condensation while being cooled to ambient temperatures; and c) chemical reactions in the atmosphere may lead to chemical species with low vapour pressure at ambient temperature. The former sources produce UFPs that are more "localised," with their concentrations decreasing with distance to their emissions sources, whereas the latter processes tend to produce UFPs that are more regionally (thus homogeneously) dispersed over an urban or rural area. UFPs sources and their impact on human exposure are discussed in the following sections.

Emission inventories suggest that motor vehicles are the primary direct emission sources of fine and UFPs to the atmosphere in urban areas. Most PNs from vehicle exhaust are in the size range of 20-130 nm for diesel engines and 20-60 nm for gasoline engines. In addition to UFP formation by direct emission, recent studies show that photochemically driven atmospheric reactions lead to the formation of low-volatility species at ambient temperature. These chemical species may form UFPs by a variety of nucleation processes. Nucleation may sometimes occur on ions and probably involves more than one species (i.e., is a multicomponent process).

There is strong evidence that sulphuric acid vapour sometimes participates in nucleation, and there is growing consensus that ammonia and water vapour are also involved. However, the atmosphere probably also contains other trace gases, including organic compounds, that either participate in the nucleation process or react in the atmosphere to form compounds that nucleate. Because they may be present at extremely low concentrations, the identity and concentrations of those gases are not yet known. Kulmala et al. (2004) have written an excellent review on this topic. A variety of different nucleation mechanisms have been proposed for the atmosphere, including binary water-sulphuric acid nucleation, ternary water-sulphuric acid-ammonia nucleation, and ion-induced nucleation.

In urban areas, nucleation events have been observed in Atlanta, Georgia, Pittsburgh, Pennsylvania, and Los Angeles. Stanier et al. (2004) showed in their study in Pittsburgh that the nucleation events are fairly well correlated with the product of ultraviolet intensity and sulfur dioxide concentration and can depend on the effective area available for condensation.

This indicates that sulphuric acid ($[H_2]S[O_4]$) is a component of the new particles. However, they noted that published correlations for nucleation by binary $[H_2]S[O_4]$-$[H_2]O$ could not explain the observed nucleation frequency and intensity, suggesting that an additional component (perhaps ammonia) is participating in the particle formation, thus supporting the notion of a ternary process. Kulmala et al. (2004) indicate that after nucleation, UFPs grow at rates ranging from 1 to 20 nm/hr (increase in physical diametre), depending on season and locale. It is of particular note that recent evidence supports the notion that the species that dominate growth may be different from the species responsible for particle formation, with particle growth being attributed to the concentration of "nonvolatile" vapours. Our current understanding of atmospheric nanoparticle processes suggests that growth of these particles to larger sizes within the UFP mode occurs by condensation of low-volatility organic species. These species are products of photochemical oxidation of volatile organic precursors on these preexisting nuclei. Recent studies by Zhang et al. (2004) showed that nucleation rates of sulphuric acid are greatly increased in the presence of organic acids (including products of atmospheric photochemical reactions), by forming unusually stable organic-sulphuric acid complexes, thereby reducing the nucleation barrier of sulphuric acid.

Exposure to UFPs

Because of the importance of traffic sources in the overall emission rates of UFPs and the resulting human exposure, it has been essential to determine UFP bahaviour after emission as they are transported away from the emission source, namely, busy roads and freeways. The most comprehensive studies of PM in the proximity of roadways were initiated in the late 1990s. Hitchins et al. (2000) measured the horizontal and vertical profiles of submicrometre particulates (16-626 nm) near a major arterial route in the urban area of Brisbane, Australia. The study found that, with the exception of measurements in close proximity to the road (~ 15 m), the horizontal profile measurements did not show statistically significant differences in fine PN concentration at ground-level distances up to 200 m away from the road. The same study examined particle size distribution and concentration in the size range of 15-200 nm and at distances from a road ranging from 15 to 375 m at two sites in Australia and found that when the wind blew directly from the road, the concentration of the fine particles and UFPs decreased to about half their maximum at a distance of 100-150 m from the road.

Almost simultaneously, Zhu et al. (2002b) measured the size distributions of PM and concentrations of gaseous co-pollutants in the proximity of a freeway mostly affected by gasoline vehicles [Interstate 405 (I-405)] in west Los Angeles. Measurements were taken at distances varying from 30 to 300 m downwind and 300 m upwind from freeway I-405. At each sampling location, concentrations of carbon monoxide (CO), black carbon (BC), and particle mass concentrations were also measured. The ranges of average concentration of CO, BC, total PN, and mass concentration at nearest location (30 m) to the freeway were 1.7-2.2 ppm, 3.4-10.0 [micro]g/[m^3], 1.3-2.0 x [10^5]/[cm^3], and 30.2-64.6 [micro]g/[m^3], respectively. For the conditions of these measurements, the decrease in concentration of CO, BC, and PN tracked each other well with increasing distance from the freeway. PN concentration (6-220 nm) decreased exponentially with downwind distance from the freeway. Zhu at al. (2002a) attributed the rapid decrease in PN concentration and change in particle size distribution to atmospheric dispersion, evapouration of volatile PM, and possibly some coagulation. Measured number concentrations tracked traffic flow well. PN concentrations on the roadway increase with vehicle speed and decrease with idling. Nearest the freeway site, three distinct ultrafine modes were observed with geometric mean diametres of 12.6 nm, 27.3 nm, and 65.3 nm. The smallest mode, with a peak concentration of 1.6 x [10^5] particles/[cm^3], disappeared at distances > 90 m from the freeway. Ultrafine PN concentration measured at 300 m downwind from the freeway was indistinguishable from upwind background concentration.

In a companion study, Zhu et al. (2002b) obtained similar data from sites in the vicinity of the 1-710 freeway in Los Angeles an area affected primarily by heavy-duty diesel traffic with the most notable difference from the 1-405 study, as BC levels were substantially more elevated in the 1-710 freeway. Similar to the study in the vicinity of the 1-405 freeway, the sharpest decrease in particle concentrations with distance from freeways was observed for the < 20 nm particles for 1-710.

Measurements of PN size distribution in the range of 11-452 nm were conducted on the side of a busy road in central London, over a period from April 1998 to August 2001 by Charron and Harrison (2003). The data were analysed to demonstrate the influences of meteorologic factors on the overall size distribution. The relationship to traffic volumes indicated that accumulation-mode particles are associated with emissions from heavy-duty traffic (mainly diesel vehicles), whereas particles in the range of 30450 nm showed a stronger association with light-duty traffic. The concentrations of both these particle size fractions decreased with increasing wind speed as a result of increased atmospheric dilution. Meteorologic parameters such as low temperatures and high relative humidity were shown to favour the formation of new particles. The relative number of particles ranging in size from 11 to 30 nm measured during the morning rush hour is strongly influenced by the prevailing temperature and humidity conditions. The strong dependence on the temperature corroborates the idea that these nanoparticles are not primarily emitted but formed during the cooling and dilution of the vehicle exhausts. By contrast no obvious relation was found between PN concentrations and relative humidity, despite the inverse correlation of temperature and relative humidity. Higher relative humidity is expected to favour homogeneous binary nucleation of sulphuric acid and water; ternary nucleation involving ammonia is expected to be independent of relative humidity, whereas the potential nucleation from organic compounds should be independent of relative humidity. The lack of a dependence on the relative humidity suggests that the binary nucleation from sulphuric acid and water is not a major factor in particle production.

The effect of season on freeway PM characteristics was also investigated in a study by Zhu et al. (2004). The decay rates of CO and BC were slightly greater in summer than in winter for both freeways, suggesting a weaker atmospheric dilution effect in winter. PN concentration in the size range of 6-12 nm is significantly higher in winter than in summer. The associated concentration in that size range decreased at a slower rate in winter than in summer. The surface area concentrations in the size range of 6-220 nm are consistently higher in summer for all sampling locations. These results suggest that wintertime conditions favour greater particle formation, possibly because of increased condensation of organic vapours coupled with decreased atmospheric mixing depth. These results are consistent with the observations made by Charron and Harrison (2003). Higher

UFP concentrations in wintertime were also observed in recent studies by Jeong et al. (2004) and Stanier et al. (2004).

Ketzel et al. (2003) performed simultaneous measurements of particle size distribution (size range, 10-700 nm) inside an urban street canyon and a nearby urban background location in Copenhagen, Denmark, in May-November 2001, attempting to separate the traffic source contribution in the street canyon from the background levels. The background concentrations were highly variable due to changing contributions from long-range transport and local sources, showing a diurnal pattern with a shift to smaller particle sizes during midday hours. As expected, higher concentrations were measured at the street site, with the average ratio of background-to-street concentration being 0.26 for nitrogen oxides (N[O_x]) and 0.35, 0.42, 0.60, and 0.64 for CO, total PN, surface, and volume, respectively. During daytime and evening hours, a maximum of particle sizes of 20-30 nm was observed in the particle size distribution, independent of the changing heavy-duty vehicle traffic during the same time interval.

All the aforementioned studies mostly, if not exclusively, reported measured aerosol size distributions with little attempt to interpret the observed results. Conversely, fundamental insight into the "road-to-ambient" evolution of PN distributions near the 1-405 and 1-710 freeways in Los Angeles in both summer and winter was developed by a multicomponent sectional aerosol dynamic model developed by Zhang et al. (2004). The model used CO as a marker of atmospheric dilution and examined the change in particle size due to evapouration and recondensation onto larger particles of volatile material because of dilution. As the vehicle exhaust leaves the tailpipe, the sharp drop in temperature and relatively high concentrations leads to significant condensation of vapour emissions, making particle composition a complex mixture. As exhaust disperses from roadways, the gas-phase concentration decreases, and some compounds may continue condensing, whereas others may begin evapourating, depending on the relative magnitude of their partial pressure and vapour pressure. Because of the atmospheric dilution process, volatile gases may evapourate from particles to achieve gas-particle equilibrium. The dynamics of volatilisation are even more pronounced for the smaller particles of the UFP range, because a higher vapour pressure is required to prevent these particles from volatilising compared with particles larger than 100 nm (in classic aerosol theory, this phenomenon is also known as the "Kelvin effect"). Smaller particles thus have to grow fast enough to minimise their Kelvin effect before the concentration of the condensing materials drops to a level that makes their growth improbable.

Zhang et al. (2004) demonstrated that condensation, evapouration, and dilution are the major mechanisms in altering aerosol size distribution, whereas coagulation and deposition play minor roles. Seasonal effects were significant with winters generally less dynamic than summers. A large number of particles grow into the > 10 nm range around 30-90 m downwind of the freeways. Beyond 90 m, some particles shrink to sizes < 10 nm, whereas others continue growing to > 100 nm as result of competition between partial pressure and vapour pressure. As a result, people who live within about 90 m of roadways are exposed to particles of very different size and chemical composition than are others who live farther away from busy streets and roadways. Particle compositions probably change dramatically as components adapt to decreasing gas-phase concentration due to dilution, so number distribution evolution is also an evolution of composition. The high concentrations of PM and gaseous co-pollutants in the proximity of freeways raise concerns on population exposure level (and its health implications) during commute.

Time spent in and near vehicles is an important route of exposure to air pollution, but few studies of UFP concentrations in vehicle-related settings have been conducted, especially inside moving vehicles. Investigators in southern California used an electric vehicle to house and power a suite of particle and gaseous pollutant measurements. Measurements were conducted on a variety of streets and freeways in Los Angeles from February through April 2003. Diesel-powered vehicles, as expected, were often a major source of high UFP count concentrations, especially when being directly followed. However, gasoline-powered vehicles were also often observed to produce comparably high UFP counts, particularly when the vehicles were older; when vehicles were accelerating hard or from a standing start, such as after waiting at a stop light; and when vehicles were driven and/or accelerated at high speeds. Because of the ubiquitous nature of gasoline-powered vehicles and the frequency of such types of driving, they may be the predominate source of in-vehicle, roadway, and near-roadway UFP concentrations. The UFP concentrations are higher than the urban background measurements by at least an order of magnitude in the freeway affected by diesel traffic (I-710) or the mostly gasoline engine freeway (I-110).

The argument that concentrations of UFPs (or, equivalently, PN concentrations) should be highly correlated with those of gaseous co-pollutants used as markers for vehicular emission, such as CO and nitrogen oxide (NO), cannot be made unilaterally. Sardar et al. (2004) performed continuous measurements of PN, PM [less than or equal to] 10 [micro]m in aerodynamic diametre (P[M_{10}]), and gaseous pollutants [CO, NO, N[O_x], and ozone ([O_3])] at urban ("source") and inland ("receptor") sites of the Los Angeles Basin for 2002 in support of the University of Southern California (USC) Children's Health Study. The degree of correlation between hourly PN and co-pollutant concentrations at each site over the entire year was generally low (i.e., Pearson correlation coefficient, $r < 0.4$). Similar analyses of the 24-hr averaged data also resulted in generally low levels of correlation. Some correlations between PN and both CO and NO were strengthened during morning rush-hour periods, indicating a common vehicular source, but when integrated over the entire 24-hr day period, they decreased dramatically. In a location near Los Angeles but not directly affected by vehicular emissions, PN concentrations were moderately to highly correlated with [O_3] in the summer, indicating that photochemical formation of PM is a more prominent source than traffic.

From the standpoint of classic nucleation theory, the photochemical production of UFPs in highly polluted areas such as Los Angeles is a somewhat surprising finding, considering that the preexisting high surface area of the urban aerosol would act as a condensational sink for precursors that would be responsible for the formation of UFPs. In addition to Los Angeles, recent studies in polluted urban areas such as Detroit, Michigan; Leipzig, Germany; and Mexico City also demonstrated a striking similarity of diurnal patterns between PNs in the smaller fractions (< 56 nm) and [O_3] only during the summer period. Pearson correlation coefficients (r) between the concentrations of UFPs in various size ranges and gaseous co-pollutants measured in the study by Sardar et al. (2005). The r-values for the smaller particles are significantly high in the summer for both Long Beach and USC sites, with r-values of 0.62-0.64 and 0.68-0.69 for particles in the size range of 0-32 and 32-56 nm, respectively. As argued above, in addition to primary emissions, photochemical reactions in the atmosphere can form new particles via nucleation or form new particle mass by condensation of low-volatility reaction products onto existing particle surfaces. It is expected that nucleation is more likely to occur during the summer with elevated photochemical activity due to larger solar zenith angles. The similarity in diurnal patterns of PNs in the < 56 nm ranges and $[O_3]$ observed for this period suggests a photochemical origin for these particles.

All the aforementioned studies indicate clearly that, besides contributions from vehicular sources, photochemical secondary formations are also a source of PM in urban atmosphere. What is important from the standpoint of human exposure is that, unlike vehicular emissions, which by nature have a strong local character whereby concentration decreases rapidly with increasing distance from their source, secondary formation of UFPs is more regional in nature, which implies a more uniform population exposure to UFPs generated by this mechanism.

Remaining Challenges in UFP Exposure Assessment

Despite the increasing concerns about the health impacts of UFPs, very little information is available on their concentrations or physical/chemical properties in places where people live and work, such as in community air, homes, schools, workplaces, restaurants, or vehicles. It is therefore essential to develop and deploy technologies that can assess the nature and extent to which people are exposed to these particles in these microenvironments. The complexity of the sources and nature of UFPs suggest that considerable characterisation efforts will be needed to either discover and/or refine our understanding of linkages between exposures and various types of health outcomes. At one extreme is the need to determine how large, spatially dispersed populations experience ultrafine exposures over prolonged periods of study. At the other extreme is the need to determine how individuals experience the hour-to-hour or minute-to-minute dynamics of ambient UFPs as they move within the aforementioned microenvironments. The analytic technologies required to characterise the physical and chemical nature of UFPs in these various microenvironments are largely unavailable or untested outside the controlled environment of the laboratory.

The information needs of health and exposure assessors suggest that various approaches should be considered to meet these needs. Air monitoring needed in support of long-term population-based or cohort studies typically must be expected to operate with very little operator intervention for periods of up to several years. Siting of such monitoring operations should be made with care to represent the overall community exposures. In time-series studies where citywide populations are often represented by central site PM monitoring, Pekkanen and Kulmala (2004) concluded that such data might be a worse proxy for human exposure to UFPs than to P[$M_{2.5}$]. In panel studies where small groups of individuals are followed for days or weeks, it may prove practical to perform enhanced monitoring of UFPs.

Enhancement might include technologies that report particle size distributions, the chemical nature of UFPs, and information on physical parameters such as shape and density. Panel studies offer the opportunity for individuals under study to carry instruments and record their locations and activities. Data from these sources have been shown to be essential when modelling overall exposures as well as in identifying the microenvironments of most concern. As with large-scale studies, careful consideration should be given to the siting of monitoring in order to provide robust data that may be related to observed effects. Although these studies offer the opportunity to determine or model exposures to individuals in the study population, it is important to develop and employ monitoring instrumentation suited to placement in homes, in classrooms, and in cars and that can be carried by children and adults. Data collected in these studies should be augmented with assessments of the activities of participants and descriptions of the microenvironments monitored.

When exposure models are developed and employed to estimate UFP exposures, it is important to consider both indoor and outdoor sources. For the purpose of estimating exposures related to time spent indoors, it is useful to gather data on infiltration of UFPs from outside sources. The extent of particle penetration into indoor environments is governed by indoor and outdoor sources, exchange rates, and particle physicochemical characteristics. Indoor particle concentrations, therefore, depend on the dynamics of the transport and fate of outdoor particles in indoor environments. Previous research in this area has focused on P[$M_{2.5}$] and P[M_{10}] properties and bahaviour. These studies indicated that concentrations of indoor particles of outdoor origin are significant. In addition, the building shell was found to be ineffective in removing infiltrating particles. Considering health implications of UFP exposure, it is important to assess particle penetration characteristics into indoor environments and the relationship between their physical and chemical properties and infiltration. Although experiments have been performed to investigate penetration properties of submicrometre particles, these laboratory-based studies have assumed that particles are spherical and rigid. Results indicated that particle size and building gap dimensions were most important factors determining particle penetration. Furthermore, real-world UFP penetration studies conducted thus far have examined infiltration properties for a limited set of conditions. For example, Long et al. (2001) evaluated penetration efficiencies in only suburban neighbourhoods. Franck et al. (2003) studied indoor and outdoor UFP size distributions at one location. Vette et al. (2001) measured indoor and outdoor particle size distributions of a single residence at urban background concentrations. However, characterisation of urban particle infiltration should consider recent studies showing that UFPs exhibit great spatial variations near sources.

Sakurai et al. (2003) studied the chemical composition and volatility of nanoparticles emitted from diesel vehicles and found that these aerosols consist of residual species, which may represent nonvolatile cores or low-volatility organic compounds as well as more volatile, smaller particles thought to be products of condensation of hot supersaturated organic vapours associated with fuel and lube oils. The volatile fraction constitutes about 90% of the total aerosol emitted by vehicles based on number concentrations. Such findings suggest that at least the volatile particles of outdoor origin can experience substantial changes and may be lost to building walls during indoor penetration.

This notion is further supported by a recent study investigating the transformation of labile ambient ammonium nitrate aerosols in indoor environments, which has shown that measured indoor concentrations were considerably lower than the values predicted based only on penetration and deposition losses. Because of the public health implications of UFPs and their spatial variations near pollutant sources such as freeways, it is therefore important to evaluate outdoor UFP size distributions, volatility properties, and penetration efficiencies into indoor environments.

In conclusion, exposure assessment issues for UFPs are complex and need to be considered before undertaking epidemiologic investigations of their health effects. For instance, because of the high spatial variability of UFPs, the use of central site concentration data alone may not reveal its relative importance compared with other PM-size fractions or compared with gaseous pollutants. Particular attention needs to be given to indoor sources and infiltration of UFPs from outside sources, as well as meteorology because of the potentially high seasonal variability in UFP PN concentrations, sources, and chemical composition, including volatility.

Nanoparticles

With the advent of nanotechnology, the prospects for using engineered nanomaterials with diametres of < 100 nm in industrial applications, medical imaging, disease diagnoses, drug delivery, cancer treatment, gene therapy, and other areas have progressed rapidly. The potential for nanoparticles (NPs) in these areas is infinite, with novel new applications constantly being explored. The possible toxic health effects of these NPs associated with human exposure are unknown. Many fine particles generally considered "nuisance dusts" are likely to acquire unique surface properties when engineered to nanosize and may exhibit toxic biological effects. Consequently, the nuisance dust may be transported to distant sites and could induce adverse health effects. In addition the beneficial uses of NPs in drug delivery, cancer treatment, and gene therapy may cause unintentional human exposure. Because of our lack of knowledge about the health effects associated with NP exposure, we have an ethical duty to take precautionary measures regarding their use. In this review we highlight the possible toxic human health effects that can result from exposure to ultrafine particles (UFPs) generated by anthropogenic activities and their cardiopulmonary outcomes. The comparability of engineered NPs to UFPs suggests that the human health effects are likely to be similar. Therefore, it is prudent to elucidate their toxicologic effect to minimise occupational and environmental exposure. Highlighting the human health outcomes caused by UFPs is not intended to give a lesser importance to either the unprecedented technologic and industrial rewards of the nanotechnology or their beneficial human uses. Key words: cons, nanoparticle toxicity, nanotechnology, pros.

The advent of nanotechnology is considered to be the biggest engineering innovation since the Industrial Revolution. Proponents of this new technology promise to reengineer the man-made world, molecule by molecule, sparking a wave of novel revolutionary commercial products from machines to medicine. This "industrial revolution" in molecular manufacturing will alter the relationship of materials so profoundly that this change may produce both positive and negative effects on health and the environment. The worldwide market for products produced using nanotechnology is estimated to reach US$1 trillion by 2015. The technologic progress during the Industrial Revolution enhanced quality of life but also resulted in a human health burden. As in the case of asbestos with its decades of long latency that still remain, there are many legitimate concerns about the unknown human health consequences of nanomaterials. Nanotechnology, now at the leading edge of rapid development with many potential human health benefits, is perceived with apprehension for potential human health risks. Enhanced strength, durability, flexibility, performance, and inimitable physical properties associated with these materials has been exploited in a multitude of industries and treatment modalities including detection of tumours, targeted drug delivery, and prognostic visual monitoring of therapy. With these applications, unprecedented avenues of exposure to nanoparticles (NPs) in humans are likely. Ambient and workplace exposures in combination with other toxic agents may cause unpredictable adverse health effects. Failure to address these imminent human health issues in a cohesive and concerted manner by industry, academia, government, environmentalists, and scientists may lead to detrimental health effects caused by exposure to NPs.

In addition to occupational exposure, direct human exposures through medicinal applications and ambient air pollution are a major concern. Inhaled NPs may evade phagocytosis, cross cell membranes, and redistribute to other sites of the body, causing systemic health effects. Therefore, the unbridled growth and use of nanotechnology in medical and human health evaluations opens society to the possibility that NPs could become the "asbestos" of the 21st century. In this review we discuss briefly some of the future human benefits of nanotechnology (pros) and emphasize possible health concerns (cons) based on the known cardiopulmonary effects of ultrafine particles (UFPs). We also discuss a limited number of studies using NPs on cellular and animal pulmonary toxicity and translocation to extrapulmonary sites. We selected the pro and con articles from the large body of literature on the basis of the number of subjects involved in epidemiologic studies and the consistency of reported studies. Studies on pro selections were based on the potential importance in biological or medicinal applications.

UFPs versus NPs

UFPs and NPs, whether anthropogenic or engineered, are similar in size with diametres < 100 nm and possess many similar characteristics. The term "UFPs" traditionally has been used to describe airborne particles with diametres < 100 nm. The term "ultrafine" is frequently used to describe nanometre-size particles that have not been intentionally produced but are the incidental products of processes involving industrial, combustion, welding, automobile, diesel, soil, and volcanic activities.

The ambient particulate matter (PM) produced from these sources contains particles in three sizes: < 0.1, 0.1-2.5, and > 2.5 [micro]m. Most of the particle mass in the ultrafine size range is < 2.5 [micro]m ([$PM_{2.5}$]), with the largest number of particles < 0.1 [micro]m. Hinds (1999) found that UFPs have longer lifetimes in the atmosphere, can be transported over thousands of kilometres, and remain suspended in air for several days.

Furthermore, UFPs with greater surface area can carry large amounts of adsorbed pollutants, oxidant gases, organic compounds, and transition metals. The greater pulmonary deposition efficiency of UFPs with larger surface area and transition metals bound to them is considered important in cardiopulmonary toxicity.

In urban industrial locations, at least half the [PM_{10}] (PM < 10 [micro]m in diametre) mass generated consists of [$PM_{2.5}$] with an average mean value of 13.4 [micro]g/[m^3]. UFPs possess a wide range of morphologic, chemical, physical, and thermodynamic properties. The UFPs emitted from different sources and geographic locations vary considerably in the types and concentrations of metal contaminants and aromatic compounds bound to surface. The primary particles emitted from the sources interact through chemical reactions in the atmosphere with oxygen, nitrogen dioxide, ozone, sulfur dioxide, and organics producing secondary particles of diverse reactivity and characteristics. The surface properties of UFPs generated at different sources and during aging of the particles are dynamically different in toxicity. Therefore, the toxicity and adverse health effects caused by UFPs are heterogeneous, depending on the source and mixed exposures of primary and secondary UFPs.

"NPs," which in general terms are defined as engineered structures with diametres of < 100 nm, are devices and systems produced by chemical and/or physical processes having specific properties not displayed in their macro-scale counterparts. Milling or grinding may also produce NPs that may or may not have properties different from the bulk materials. The term "NPs" in this review is used, therefore, to differentiate engineered particles with a diametre of < 100 nm that are different from incidental UFPs.

Pros: Applications in Biology and Medicine

Imaging and diagnosis: Molecular imaging is an important discipline in biology and medicine with ability to detect, quantify, and display molecular and cellular changes that happen in vitro and in vivo. Fluorescent biological probes are used conventionally in biology because of their inert qualities and their ability to interact without loss of sensitivity in a variety of cellular reactions. However, there are intrinsic limitations with several organic dyes. The dynamic range of NPs, with diametres of < 100 nm, as probes attached to molecules of peptides, antibodies, or nucleic acids for the detection of cellular reaction products makes them ideal tools for display and quantification of molecular reactions in vivo. Such NP-based probes have high levels of brightness, photostability, and absorption coefficients across a wide spectral range. Their abilities to monitor ultrastructural interactions on a continuum make them ideal for applications in biology and disease. Furthermore, the potential for coating the NPs with antibodies, collagen, and other micromolecules makes them biocompatible for detection and diagnosis.

An increasing number of studies in diagnosis and detection have been published, and we describe a select few here. In a study using mouse fibroblasts, Bruchez et al. (1998) showed that NP-based fluorescent rabelling was better than conventional fluorophores. Wu et al. (2003) observed that quantum dot-based immunofluorescent rabelling of the cancer marker Her2 was more efficient than conventional fluorophores in rabelling different target cell-surface receptors, cytoskeleton, nuclear antigens, and other intracellular organelles. They also demonstrated that bioconjugated colloidal quantum dots were valuable in cell rabelling, cell tracking, DNA detection, and in vivo imaging. Zhang et al. (2002) showed the surface modification of superparamagnetite NPs with ethylene glycol and folic acid was effective in facilitating phagocytosis by cancer cells for potential cancer therapy and diagnosis. Gao et al. (2004) reported imaging and cancer targeting based on semiconductor quantum dots in animal studies in vivo. In control studies Gao et al. observed the uptake, retention, and distribution of quantum dots primarily in the liver, spleen, brain, heart, kidney, and lung in decreasing order. In nude mice growing human prostate cancer xenograft, quantum dots accumulated specifically at cancer targets showing bright orange red colour. Drug delivery: Site-specific-targeted drug delivery is important in the therapeutic modulation of effective drug dose and disease control. Targeted encapsulated drug delivery using NPs is more effective for improved bioavailability, minimal side effects, decreased toxicity to other organs, and is less costly. NP-based drug delivery is feasible in hydrophobic and hydrophilic states through variable routes of administration, including oral, vascular, and inhalation.

In drug delivery, several approaches are currently being tested for better site-specific delivery of an effective dose using liposomes, polymeric micelles, dendrimeres, ceramic NPs, iron oxide, proteins, covalent binding, adsorption, conjugation, and encapsulation methods. Extended circulation of liposomes with entrapped doxorubicin was reported to be 300-fold more effective, with better pharmacokinetic ability than free doxorubicin in the treatment of Kaposi's sarcoma and metastatic cancer. Kumar et al. (2004) reported that NP surfaces modified with cationic chitosan were efficient for drug delivery both in vitro and in vivo. Gelperina et al. (2005) reported that in chemotherapy treatment for tuberculosis, NP-based drug delivery improved drug bioavailability, reduced dose frequency, and overcame the nonadherence problem encountered in the control of tuberculosis epidemics.

Anticancer therapy: Conventional anticancer treatments are nonspecific to target killing of tumour cells, may induce severe systemic toxicity, and produce drug resistant phenotypic growth. An exciting potential use of nanotechnology in cancer treatments is the exploration of tumour-specific thermal scalpels to heat and burn tumours. O'Neal et al. (2004) observed in mice that selective photothermal ablation of tumour using near infrared-absorbing polyethylene-coated gold nanoshells of 130 nm inhibited tumour growth and enhanced survival of animals for up to 90 days compared with controls. Perkel (2004) also reported that antibody-coated magnetic iron NPs were effective to heat and literally cook the tumours. In similar work performed in athymic mice using antibody-coated iron NPs, DeNardo et al. (2005) showed specific targeted binding to tumours and tumour necrosis within 24 hr after therapy with better response. The efficacy of different antibodies conjugated to NPs, including transferrin and epidermal growth factor receptor, was examined in animal studies. In cancer therapy, enzyme-mediated liposome destabilisation and specific phospholipase [A_2] activation with synergistic membrane perturbing and permeability were reported to be more effective.

Gene therapy: Attempts to cure genetic diseases by transfer of somatic cells transfected with normal genes gained popularity in the last two decades. In gene therapy a normal gene is inserted in place of an abnormal disease-causing gene using a carrier molecule. Conventional uses of viral vectors are associated with adverse immunologic, inflammatory reactions, and diseases in the host. In this regard Gopalan et al. (2004) found NP-based gene therapy to be effective in systemic gene treatment of lung cancer using a novel tumour suppressor gene, FUS1.

Chitosan, a polymer long used in gene therapy, was reported to have increased transfection efficiency and decreased cytotoxicity. Oral gene delivery in BALB/C mice using poly-L-lysine modified silica NPs has shown success with the distribution of particles throughout the intestinal mucous cells with limited cytotoxicity. Dufes et al. (2005) reported gene therapy by intravenous administration of NP-based vector systems using tumour necrosis factor (TNF)-[α] expression plasmid and found increased transgene expression and long-term survival of rats with no toxicity. Kaul and Amiji (2005) observed that tumour-targeted gene delivery using polyethylene glycol-modified gelatin NPs was highly effective, biocompatible, biodegradable, and long circulating for systemic delivery to solid tumours. Recent in vitro work with breast cancer cells has shown the potential efficacy of NP-mediated gene delivery of the wild-type p53 gene. Cancer cells exposed to this NPs-based gene delivery showed an increased and sustained antiproliferative activity not seen in cells exposed to vector alone. Bharali et al. (2005) reported that a nonviral vector for in vivo gene delivery and fluorescent visualisation of transfection using organically modified silica NPs has promising success for targeted brain therapy. The efficacy of NP-based transfection exceeded viral vector-based gene delivery, and in vivo optical imaging provided efficient and continual monitoring, retention, and viability of transfected cells. From these studies, it is apparent that nanotechnology will profoundly affect human health through advances in medicine, science, and industry. The potential human benefits of nanotechnology are innumerable and include many aspects of human life with wide a variety of products.

Cons: Morbidity and Mortality Due to Cardiovascular Effects

Caveat. Since the beginning of the Industrial Revolution, anthropogenic sources of human exposure have increased dramatically, and based on a temporal correlation, the high concentrations of ambient air pollution and increases in morbidity and mortality were well established by several epidemiologic studies. However, these epidemiologic data are not supported by a direct cause and effect relationship.

Numerous epidemiologic investigations have shown a direct credible relationship between ambient air particulate pollution and a consistent association with increased health effects specifically attributed to cardiovascular diseases. During the last few decades there has been a continued increase in the morbidity and mortality among adults and susceptible populations attributed to air pollution in industrialised and developing countries. The concentration response relationship between [$PM_{2.5}$] and daily deaths has been reported to cause 100,000 deaths annually in the United States. In a recent comprehensive review of epidemiologic studies, Delfino et al. (2005) showed clearly the pathophysiologic changes associated with exposure to UFPs changes that induce cardiovascular diseases.

A strong association of ambient particulate air pollution as a predictor of mortality and morbidity of adults in six polluted and less polluted U.S. cities was well documented in early two epidemiologic studies. In a subsequent study the fine particulate burden was further linked to increased cardiovascular mortality and morbidity with physiologic correlates. In this study an increased heart rate was reported to be associated with increased exposure to airborne ambient particulates. Exposure to ambient air pollution was also shown to be associated with an increase in blood pressure and decreased heart rate variability with no apparent changes in oxygen saturation. Peters et al. (2000, 2001) showed that elevated levels of air pollution are associated with increasing incidence of life-threatening arrhythmia and triggering of myocardial infarction. They also showed that exposure to increased levels of air pollution for short durations of [greater than or equal to] 2 hr triggered myocardial infarction. In a fine particulate air pollution and mortality study in 20 U.S. cities, a 0.68% increase in relative death rate from cardiovascular and respiratory causes was reported for each increase in the [PM_{10}] level of micrograms per cubic metre. Epidemiologic and pathophysiologic evidence supported the link of fine particulate air pollution to cause-specific cardiovascular mortality to diseases such as pulmonary and systemic inflammation, accelerated atherosclerosis, and altered cardiac autonomic function.

In a 16-year followup of 500,000 adults in the Cancer Prevention Study II, Dockery et al. (2005) reported that a 10 [micro]g/[m^3] increase in [$PM_{2.5}$] was associated with an 8-18% increase in mortality due to ischemic heart disease, dysrhythmias, heart failure, and cardiac arrest.

Exposure for a short period of 3 days to increased ambient air pollution was also reported as a major factor for the increased disease outcomes. In recently published population-based studies, it was also reported that short-term exposures to ambient [$PM_{2.5}$] and [O_3] are associated with cardiac autonomic dysfunction in older adults who had histories of cardiac and other diseases, showing an effect modification. Regional differences associated with [$PM_{2.5}$] and cardiovascular respiratory risks in the eastern, northwestern, and southwestern United States compared with other regions of the country were also reported recently. The potential for UFPs to exacerbate preexisting cardiovascular risks is implicated in this effect. Peters et al. (1997) in a retrospective study on plasma viscosity during an air pollution episode reported an increase suggestive of a link to mortality.

The hypothesis that translocation of inhaled UFPs directly to systemic circulation causes a direct cardiovascular effect in individuals at increased risk was tested in a study on healthy subjects. In this study inhaled ultrafine carbon particles rabelled with technetium (Tc) were shown to pass rapidly into the systemic circulation of healthy male nonsmokers. However, this finding has been refuted by several other studies as a methodologic overestimation of soluble [^{99}Tc] pertechnetate. In a recent study Mills et al. (2006) showed conclusively that the majority of [^{99}Tc]-rabelled carbon nanoparticles remained in the lung for up to 6 hr after inhalation.

Three mechanisms have been proposed to explain the events that may lead to cardiopulmonary morbidity and mortality in populations exposed to fine ambient air pollution. The first hypothesis is that the fine PM are able to stimulate the neurons in the lung affecting the central nervous system and cardiovascular autonomic function. The second proposes that inhaled fine PM gains direct access to the systemic circulation and reaches target organs, thereby triggering inflammation, secretion of cytokines, reactive oxygen species (ROS), C-reactive proteins, and inducing cardiac events. The third hypothesis proposes that inhaled fine PM triggers an acute inflammatory response in the lung, thus stimulating the secretion of cytokines, chemokines, ROS, and transcription factors. The cascade of subsequent events and inflammation plays a key role in the activation of mitogen-activating protein kinase (MAPK), redox-sensitive transcription factors, nuclear factor kappa B (NF-[kappa]B), and activating protein-1 (AP-1), thereby promoting pulmonary inflammation that leads to cardiac events. There is a strong link supporting the relationship between inflammation and coronary heart disease because inflammation is associated directly with atherosclerosis. Furthermore, results from animal and cellular studies support credible mechanistic pathways induced by inflammatory responses that lead to systemic disease. In this respect, studies using genetically susceptible mice exposed to long-term air pollution PM showed acceleration of atherosclerosis and vascular inflammation, thereby supporting the indirect link between inflammation-induced developments of atherosclerosis. Evidence from human studies also supports this link to air pollution PM and development of cardiac responses that lead to atherosclerosis.

Several epidemiologic studies present the argument that ambient particles in the [$PM_{2.5}$] range can be correlated with adverse health effects. Characterisation of human exposures to UFPs in these studies has been a limiting factor. Most of these studies used available indirect monitoring data to correlate adverse health effects without information on the proportions of UFPs or its specific toxic components. Consequently, criticisms have been raised about the possible nonspecific causal correlation for the observed associations. The lack of toxicologic evidence supporting these epidemiologic studies is answered partly by cellular and animal studies.

Animal experimental studies using concentrated ambient particles (CAPs) suggest that pulmonary vasculature is an important target for ambient air particle toxicity, as these particles exacerbate myocardial ischemia and induce coronary artery occlusion. Batalha et al. (2002) found that short-term exposure to CAPs induced vasoconstriction of small pulmonary arteries in normal rats and rats with chronic bronchitis. In dogs with coronary artery occlusion, Wellenius et al. (2003) observed that inhalation of CAPs resulted in the exacerbation of myocardial ischemia. The effect of UFPs in inducing experimental thrombosis in animal models supports the effect of particle size on the development of vascular thrombosis and pulmonary inflammations. This ability of UFPs to evade phagocytosis and enter systemic circulation to reach extrapulmonary sites may be a relevant mechanism involved in cardiovascular mortality and morbidity, but it is unclear at this time. Seaton et al. (1995) hypothesized that a) UFPs reaching the cardiovascular system may induce coagulation, thrombosis, or other impairments or b) the persistent inflammation from UFPs in the lung promotes the release of mediators and cytokines, thus triggering cardiopulmonary events that lead to increased morbidity or mortality. In support of this hypothesis, Delfino et al. (2005) reported that patients with coronary heart disease had increased levels of inflammatory cytokines such as interleukin (IL)-1[β], IL-6, TNF-[α], C-reactive protein, and fibrinogen, compared with unaffected people.

This phenomenon was investigated by Nurkiewicz et al. (2004). Using an animal model, they showed the potential involvement of the systemic circulation caused by exposure to residual fly ash (ROFA < 2 [micro]M), a surrogate for UFPs, or titanium dioxide (Ti[O_2] < 1 [micro]M) particles. They demonstrated that exposure to fine particles can impair systemic microvascular functional changes independent of any detectable pulmonary inflammation. Exposure to fine PM was associated with an influx of leukocytes in trapezius muscle venules of rats, demonstrating an endothelium-dependent arteriolar dilation and impairment. This was supported further by an increase in systemic blood pressure and a failure of microvessels to respond to intraluminal vasodilators.

These impairments may be the contributing factors involved in compromising the cardiovascular system, thereby leading to exacerbations and increased risk for heart attack in UFP-exposed populations. A recent study by Li et al. (2005a) showed that other mechanistic events induced by UFPs produced vasoconstriction by enhancing MAPK signalling via angiotensin type 1 receptor activation. Urban UFPs produced a time- and dose-dependent increase in phosphorylation of extracellular signal regulated kinase (ERK) 1/ERK2 and p38 MAPK. Copper and vanadium, two common metal contaminants in UFPs, also induced this activation of the local rennin-angiotensin system that plays an important role in cardiovascular effects. The water-soluble fraction containing copper and vanadium also induced phosphorylation of ERK1/ERK2 and p38 MAPK. In vivo support of these in vitro studies on the role of oxidative stress was demonstrated in an animal model compromised by pretreatment with dimethyl urea. The molecular mechanisms promoted by ROFA-induced oxidative stress resulted in the activation of MAPK, inflammatory cytokines TNF-[α] and IL-6, and inflammatory protein MIP-2 (macrophage inflammatory protein 2) and provide insight into an inflammation-dependent triggering of events in the lung.

Cons: Pulmonary Morbidity and Mortality

The lung is the major target of ambient air pollution and the relationship between increased ambient air pollution and adverse health effects in children, individuals with asthma, and vulnerable adults is well documented. Particle size, surface area, and chemical composition all play a role in the health risks posed by PM. Increased respiratory symptoms, increased hospitalisation, decreased lung function, increased respiratory infections, altered mucociliary clearance, chronic obstructive pulmonary disease (COPD), and increased mortality are some of the major documented adverse health effects caused by exposure to ambient air pollution.

In provoking pulmonary health effects, exacerbation caused by pulmonary inflammation appears to play a major role. In susceptible individuals with asthma and COPD patients, exacerbation appears to be the important molecular mechanism by which UFPs exert their toxicity. Conversely, long-term health effects such as pneumoconiosis and cancer, which have subtle early detectable symptoms, remain difficult to establish because of the long latency of these diseases. Diseases exacerbated by acute inflammation such as asthma and COPD are well documented with corresponding fluctuations in ambient air pollution. It is reasonable to expect that different molecular mechanisms may be involved in the genesis of cardiovascular and pulmonary diseases. Experimental studies have consistently documented that exposure to UFPs and NPs is more inflammatory to the lung, and a fraction of the inhaled UFPs is translocated to different extrapulmonary sites of blood, liver, heart, spleen, and brain. The extrapulmonary translocation is variable depending on particle size, chemical characteristics, and surface features. A recent study by Geiser et al. (2005) shows that inhaled $TiO[_2]$ UFPs cross cellular membranes by nonphagocytic mechanisms in the lungs and were found in capillaries.

Experimental cellular studies: Experimental studies using laboratory-generated UFPs and airborne CAPs have shown consistently higher pulmonary inflammatory and toxicity responses to UFPs. It is believed that UFPs provoke an increased oxidative stress because of their greater surface areas that allow them to interact with more cellular structures and different types of transition metals often associated with these particles. The importance of the surface area within a narrow size range from 10 to 50 nm was recently demonstrated using acute lung inflammation as an end point after exposure to six different UFPs. Interactions between UFPs and associated transition metals were reported to have a synergistic mechanism in ROS generation and inflammation. Differences in size and composition of UFPs compared with fine and larger particles were well correlated in studies of their uptake by different cell systems and their ability to induce oxidative stress. UFPs were shown to be the most potent inducers of oxidative stress in macrophages and epithelial cells by inducing heme oxygenase-1 and depleting intracellular glutathione. Oxidative stress induced by UFPs was also reported to be involved in the activation of MAPKs, which leads to the intracellular signalling of gene expression, and in the activation of AP-1 and NF-[kappa]B, which are important in the expression of proinflammatory genes and cytokines, including adhesion molecules.

Reactive oxygen species: In vitro and in vivo studies using CAPs and laboratory-made UFPs with different chemical composition and particle sizes have shown that ROS production is a major contributing factor in inflammation and toxicity. Investigators attribute the ability of UFPs to cause lung injury and disease to the larger surface area, smaller size, and metal contaminants. In addition NP-induced generation of ROS leading to oxidative stress, activation of signalling pathways, and apoptosis are considered new insights into the development of pulmonary and other diseases. This was illustrated in a hierarchical oxidative stress model that showed the correlation of oxidative stress and corresponding changes in cellular responses induced by UFPs.

Among the pathways implicated for the increased cardiopulmonary mortality and morbidity, the oxidant-dependent proinflammatory mechanisms are considered important based on in vitro and in vivo studies. The oxidative potential of UFPs collected from divergent sources such as natural dust, oil fly ash, coal fly ash, and ambient air was attributed primarily to their metal composition. Several metals identified in UFPs can generate [OH] radicals by Fenton-like reactions directly or after cellular reductions. It was also shown that both water-soluble and insoluble components of the UFPs have oxidant generation potential. In a study of healthy and pathogen-challenged animals, Antonini et al. (2002) observed that ROFA exposure resulted in severe lung damage and inflammation, thereby implicating altered oxidative stress in susceptible exposed animals.

Pulmonary pathologic response: In vivo studies using CAPs or laboratory-generated UFPs also showed significantly greater inflammatory and toxic pulmonary responses in animals, depending on fine particle size and chemical composition. Oberdorster et al. (1994) observed that ultrafine Ti[O_2] instilled into rats and mice was more proinflammatory than fine Ti[O_2], as Ti[O_2] induced a neutronphilic influx into the lungs. Oberdorster et al. (1994) reported that particle surface chemistry is equally or more important in inflammation and acute toxicity. Freshly generated polytetrafluethylene (PTFE) fume containing UFPs < 26 nm induced hemorrhagic pulmonary inflammation and death after rats were exposed to 10-30 [micro]g of dust, and aging of UFPs resulted in the loss of surface reactivity and toxicity.

Rats are often considered to be a sensitive and exaggerated-response animal model for particle-induced pulmonary investigations. Therefore, investigators in two recent independent studies used mice that were exposed to single-wall carbon nanotubes (SWCNT) and observed significant pulmonary pathologic changes with small as well as higher doses. In the study by Lam et al. (2004), all doses (3.3-16.6 mg/kg body weight) induced granulomatous lesions with persistent inflammation up to 90 days. Shvedova et al. (2005), using SWCNTs with minimal impurities at a dose of 10-40 [micro]g/mouse, also found robust acute inflammatory response with the onset of pulmonary fibrosis associated with decreases in pulmonary function.

A comprehensive study conducted using the sensitive-response animal model (rat) in parallel with toxicity assessment showed controversial results. Rats exposed to SWCNT, silica, or carbonyl iron as positive and negative controls, respectively, showed surprisingly contradicting results. They reported pulmonary granuloma formation in the absence of toxicity, inflammatory influx, and possible regression of granulomas over time. The spectrum of toxicologic responses that is so well documented with particulate exposure in several animals and humans is inconsistent and doubtful with respect to the data presented by Warheit et al. (2004) when one compares the dose to which animals were exposed per gram body weight. Lam et al. (2004) and Shvedova et al. (2005) exposed mice to doses smaller than those used by Warheit et al. (2004) in the sensitive animal model, and reported findings of multifocal granulomas of no physiologic significance. The relevance of this speculative interpretation is unfounded.

Cons: Translocation and Toxicity to Other Organs

In the past, cardiovascular, neurologic, and excretory systems have not been considered secondary targets in inhalation toxicology and pathobiology. However, in recent years many animal and human studies have shown the translocation of UFPs to extrapulmonary sites such as systemic circulation, liver, heart, and brain. Although currently the process of UFP translocation is poorly understood, these preliminary studies provide consistent toxicologic backup for the hypothesis that UFPs are translocated to other organs, including heart, thus, playing a role in triggering and/or promoting cardiovascular morbidity and mortality. Several research studies involving UFPs have demonstrated the enhanced ability of these particles to penetrate more deeply into the lung interstitium than larger particles and evade clearance. This ability of UFPs to evade clearance promotes longer retention time in pulmonary interstitium, thus increasing the potential for translocation to extrapulmonary sites to exert effects. In an inhalation exposure study of rats, Oberdorster et al. (2002) reported that a significant amount of [^{13}C] accumulated in liver within 30 min postinhalation with a 5-fold increase in 1 day. They also observed that after inhalation, UFPs were transported to olfactory nerves at a speed of 2.5 mm/hr. From the available evidence, Hoet et al. (2004) concluded that phagocytosis by alveolar macrophages and endocytosis by the epithelial and endothelial cells are the important routes for translocation of UFPs to systemic circulation and then to other extrapulmonary sites.

Neuronal translocation: The potential for neuronal uptake and translocation of inhaled particulates and pathogens to the brain, which has been reported in several studies, was reviewed in detail by Oberdorster et al. (2005b). According to studies cited in their review, the olfactory nerve is the most viable pathway for the transport of particles intranasally inhaled because of the close proximity of the olfactory mucosa and bulb. In studies using whole-body inhalation exposure of rats to ultrafine carbon black, extrapulmonary translocation through the olfactory nerve was reported to be a viable mechanism. Whether these NPs that are transported to the brain cause cell injury or toxicity to the brain is not known.

Dermal exposure and translocation: The human skin is the largest organ in the body, protecting against the environment with a surface area of nearly 18,000 [cm^2]. Polar and nonpolar materials can permeate across the stratum corneum via a paracellular route. Photomechanical waves have been shown to enhance permeability of the stratum corneum in vivo, inducing expansion of the lacunar spaces, which leads to the formation of transient channels to facilitate the transport of macromolecules into the viable epidermis. A metabolic intervention to enhance effective transdermal drug delivery is reported to be highly effective in through permeabilised stratum corneum.

Penetration of particles > 1 [micro]m is limited through healthy skin except in areas that have been scratched, injured, or mechanically stretched. In conjunction with physical activity, Tinkle et al. (2003) demonstrated in an animal model that topically applied 0.5- and 1.0-[micro]m beryllium particles penetrate the stratum corneum and develop hapten-specific, cell-mediated immune response. Penetration of Ti[O_2] microparticles contained in sunscreen into the stratum corneum and follicular orifice of hair has been reported. Particles reaching the dermis can be transported to the lymphatic system by macrophages and dendritic cells. Although there are no well-documented studies on NP penetration and transmigration to other distant organs, one can assume from the results of reported studies using beryllium particles that it is more likely a viable route of entry in workers involved in mechanical or strenuous activities.

In a cytotoxicity study of human keratinocytes in culture, Shvedova et al. (2003) demonstrated the potential of SWCNT exposure to induce ROS generation, which results in cytotoxicity, lipid peroxidation, antioxidant depletion, and loss of cell viability associated with ultrastructural and pathologic changes. Their studies concluded that exposure to unrefined SWCNT can result in accelerated oxidative stress and toxic manifestations in workers. On the basis of current available toxicologic studies and limited human data, we have developed a schema of potential interactions of UFP transportation, and suspected sequence of events that may lead to cardiovascular, pulmonary and other organ involvement.

Advances in nanomedicine offer the possibility of new and intriguing opportunities in NP-based early detection, diagnosis, and treatment of diseases. Commercial development of nanotechnology and its many applications may unleash a spectrum of human health hazards that at the present time can only be speculated about without any detailed understanding of the toxic nature of NPs. The review of literature in the fields of toxicology and the possible human health effects of UFPs and NPs provide only a glimpse of some toxic paradigms that compel us to weigh the adverse effects against the beneficial effects. Because we know little about the toxic health hazards of NPs in vivo and in vitro, pharmacokinetic and toxicologic studies are mandatory before large-scale industrial production and use are implemented. In this regard the U.S. Environmental Protection Agency, the International Life Sciences Institute Research Foundation, and the Risk Science Institute convened working groups comprising experts in the fields of nanotechnology from academia and government to develop new toxicity screening, reporting, and hazard identification of engineered nanomaterials.

Although at this time, the benefits of nanotechnology dominate our thinking, the potential for undesirable human health outcomes should not be overlooked. Consistently large numbers of studies have reported associations between UFP exposure and morbidity in elderly and compromised individuals. Furthermore, recent studies also emphasize the impact of day-to-day variations in particle concentrations and exposures for short periods as important factors in cardiac events in predisposed population. Therefore, there is reason to suspect that NPs with size and surface characteristics similar to UFPs are likely to cause diseases some with a long latency. With widespread industrialisation of nanotechnology, there is the potential for ambient air pollution and a conceivable threat to the general population.

Doubtless, nanotechnology will have a profound impact on a wide range of applications and therefore on many aspects of human life, including environmental decontamination, water purification, cheaper electricity, and better disease treatment modalities. One major challenge facing industry and government is the lack of information on the possible adverse health effects caused by exposure to different nanomaterials. Development of safety guidelines by government for the nanotechnology industries, including manufacturing, monitoring of worker exposure, ambient release of NPs, and risk evaluations, is mandatory to promote nanotechnology for its economic incentives and medicinal applications.

Surfactant on the Retention of Man-made Vitreous Fibres (MMVF 10a)

Inhaled and deposited man-made vitreous fibres (MMVF) 10a (low-fluorine preparation of Schuller 901 insulation glass) were studied by electron microscopy in hamster lungs, fixed by intravascular perfusion within 23 [+ or -] 2 min (SD) of the initial inhalation. We found fibres on the surfaces of conducting airways and alveoli. In the airways, 89% of the fibres were totally and 11% partially covered by lining-layer material. In the alveoli, 32% of the fibres were totally submersed; others touched the alveolar wall, stuck at one end, bridging the airspace. Studies in a surface balance showed that fibres were immersed into the aqueous subphase by approximately 50% at film surface tensions of 20-25 mJ/[m^2] and were submersed (totally immersed; i.e., totally surrounded by fluid) at approximately 10 mJ/[m^2]. Fibres were also found to be phagocytosed by macrophages. We found a substantial number of particle profiles within alveolar blood capillaries. Fibre length and alveolar geometry appear to be important limiting factors for the submersion of vitreous fibres into the lungs' surface lining layer.

Man-made vitreous fibres (MMVF) are inorganic amorphous fibres made from glass, rock, clay, slag, or pure oxide raw material. They are used primarily for thermal and acoustic insulation. The fact that asbestos fibres are carcinogenic has led to the fear that the inhalation of vitreous fibres, which have similar size and shape, poses a serious risk to human health. The mechanisms by which fibres induce lung disease are still not completely understood. The potential toxicity of a given type of fibre is determined by dimension, durability, and fibre dose in lungs. Broad ranges of inhalation studies with different fibre types have been performed and have been extensively reviewed by Hesterberg and Hart (2001). A strong relationship between biopersistence of the fibre and pathogenicity in lungs has been reported. Most of the tested vitreous fibres were less biopersistent and caused less pathogenicity than asbestos fibres.

Despite numerous studies with reference to pathogenicity, little is known about how inhaled vitreous fibres interact with the inner surface of the lungs immediately after their deposition. Knowledge about the mechanisms of fibre retention is important with respect to the clearance of vitreous fibres and their possible risk to human health. The surface-lining layer in lungs consists of an aqueous phase and a surfactant film at the air-liquid interface. The continuous aqueous layer adjacent to the epithelial cells has a relatively low viscosity. The existence, thickness, and continuity of a gel phase above this periciliary layer in the various airway compartments and species have been disputed. In small airways, a thick mucus layer has never been found. The surfactant film at the air-liquid interface covers the aqueous phase and it is based on all available evidence continuous between the alveoli and central airways.

Studies in vitro have shown that the surfactant promotes the displacement of polystyrene particles from air into an aqueous subphase, whereby the extent of immersion depends on the surface tension of the surfactant film: the lower the surface tension, the greater the immersion. Furthermore, there is evidence from animal studies for the displacement of a variety of spherical particles in lungs. The influence of the shape of particles on their immersion into the aqueous lining layer is largely unknown.

A surface tension of 32 mJ/[m^2] was measured in the trachea of horses, and values for expiration in alveoli were below 2 mJ/[m^2]. Very recently, evidence for surface tension values below 15 mJ/[m^2] in the intrapulmonary conducting airways was obtained from particle deposition studies in hamsters. It is not known how different surface tension values would influence the displacement of fibres into the lining layer. The immersion of fibres into the surface-lining layer and their displacement toward the epithelium by surfactant may have consequences for fibre dissolution and/or for the clearance pathway of fibres. To evaluate the distribution of fibres within lungs immediately after their deposition, inhaled and deposited MMVF 10a were studied in hamsters by electron microscopy. The immersion of fibres, promoted by a surfactant film formed on an aqueous substrate, was assessed in a surface balance by light microscopy.

Fibres: MMVF 10a were kindly provided by Philippe Thevenaz (Research and Consulting Company, Fullinsdorf, Switzerland). They have a smooth surface and are mostly straight cylinders.

To estimate the surface free energy of the glass fibres used in this study, we measured the contact angle of dibutyl phthalate droplets (surface tension, 34.5 mJ/[m^2] at 22°C) placed on glass slides cleaned with chromic acid at different time points after exposure to laboratory air. The contact angle of dibutyl phthalate droplets on glass was compared with that obtained on a polystyrene substrate with the same fluid. In addition, the solid-vapour free energy was estimated by using the equation of state approach. The equation of state helps determine the solid-vapour interfacial energy from the measured contact angle that a test fluid droplet of a known surface tension makes on a solid substrate.

Animals, aerosol generation, and inhalation: Inhaled and deposited fibres were studied in male Syrian Golden hamsters [n = 4; AURA, HanIbm: AURA (specific-pathogen free); Biological Research Laboratories Ltd., Fullinsdorf, Switzerland] weighing 132-163 g. The animals, aerosol generation, and inhalation have been described in detail previously. Briefly, aerosols were generated by jet nebulisation of droplets of suspensions of MMVF 10a fibres (2 mg/mL in 100% ethanol) into a particle-free air stream. The fibres were then dried, charge equilibrated, and concentrated before being injected into a reservoir and transferred to the inhalation/exhalation channel. The deeply anesthetized and completely relaxed hamsters inhaled the aerosol directly from the reservoir through an intratracheal cannula during continuous negative-pressure ventilation. They were ventilated with a tidal volume ([V_T]) of 0.95 [+ or -] 0.06 mL (mean [+ or -] SD) maintained at 70-80% of total lung capacity and a breathing frequency (f) of 61 [+ or -] 1 breath/min (mean [+ or -] SD), which equals slow, deep breathing. The breathing parameters and the number of inhaled and exhaled fibres were measured by pneumotachography and laser-light scattering photometry. Because the fibres used in this study were heterogeneous in size, number estimates are of directional value only. Aerosols were inhaled for 5-12 min, until a sufficient number of fibres for microscopic analysis were deposited. Immediately thereafter, the lungs were prepared for intravascular perfusion fixation.

We collected aerosol samples on filters and analysed the fibre size distribution. The average length of 200 fibres measured by light microscopy was 16 [+ or -] 13 [micro]m (mean [+ or -] SD; range, 1-66 [micro]m); the average diametre was 1.1 [+ or -] 0.2 [micro]m (range, 0.6-1.7 [micro]m). The average length-to-diametre ratio was 15.6 [+ or -] 13.3 (range, 1.1-64.0). Twenty fibres had a length-to-diametre ratio of < 3; hence, the particulate fraction of MMVF 10a in the aerosol was 10%.

Lung Fixation

Whole lungs can be preserved for morphologic analysis by the introduction of a suitable fixative into the airways of a collapsed lung or by delivery of the fixatives to the lungs via the blood vessels. The instillation of aqueous fixative solutions into the airways is not suitable for the investigation of the inner surface of the lungs because the surface-lining layer and all its cellular and acellular constituents are dislocated from their native positions and transferred to other regions of the organ. Instillation of airways with nonpolar fixative solutions (1% osmium tetroxide dissolved in perfluorocarbon) has been demonstrated to preserve the aqueous lining layer and especially also the surfactant film at the air-liquid interface. However, it is not known whether material that is not submersed in the surface-lining layer may be dislocated by this procedure.

Because it was the aim of this study to investigate the location of deposited fibres with respect to the inner surface of the lungs, the fixative solutions were applied by vascular perfusion via the pulmonary artery, which allows the airways and alveoli to remain in their natural air-filled state. Sequential intravascular perfusion of buffered 2.5% glutaraldehyde, 1% osmium tetroxide, and 0.5% uranyl acetate has been shown to well preserve the surface-lining layer with its phagocytic cells and the surfactant film at the air-liquid interface of intrapulmonary conducting airways and alveoli in laboratory animals. All lungs were fixed within 23 [+ or -] 2 min of the initial inhalation. The short time between inhalation and lung fixation allows the investigation of fibres as close as possible to where they were deposited. After fixation, the lungs were removed from the thorax and split into left lung, right lung, and accessory lobe.

Tissue Processing for Microscopic Analysis

The lung lobes were cut perpendicularly to the longitudinal axis into series of thick (3 mm) and thin (1.5 mm) slices, with random start. The thin slices were used for electron microscopy, and the thicker ones for other purposes. A total of 49 thin slices were sampled for transmission electron microscopy (TEM) and further cut into cubes of about 1 mm side length. These tissue blocks (n = 144) were dehydrated in ethanol and embedded in Epon. Ultrathin sections, stained with uranyl acetate and lead citrate, were examined with a Philips 300 transmission electron microscope operating at 60 kV. For scanning electron microscopic (SEM) analysis, a total of 15 thin slices were sampled, dehydrated in ethanol, critical-point-dried, and sputter-coated with platinum. They were examined with a Philips XL 30-FEG scanning electron microscope (Philips) operating at 10 kV. Length and diametre of all registered fibres were measured by SEM.

Displacement of MMVF 10a in the Langmuir-Wilhelmy Balance

The displacement of MMVF 10a fibres into an aqueous substrate by a surfactant film was investigated with a monolayer of 1,2-dipalmitoyl-sn-3-glycerophosphatidylcholine in a Langmuir-Wilhelmy surface balance (our own design). The DPPC monolayer was spread from a solution of 2 mg/mL in ethanol onto an aqueous substrate of a solution of 0.9% NaCl and 40% sucrose (density, 1.25 g/mL). The monolayer was spread within an area of 25 [cm^2] enclosed by a Teflon ribbon supported by a rhombic frame. Our objective was to assess the displacement of glass fibres at surface tensions between 30 and 5 mJ/[m^2]. Pulling the rhombic frame to an elongated shape reduced the surface tension of the DPPC film, and the surface tension force was measured by a platinum Wilhelmy plate connected to an electrobalance. After adjustment of the monolayer surface tension to a preset value, the glass dish was placed onto a microscope stage and fibres were blown onto the monolayer. The fibres were then observed and photographed with a Nikon Optiphot metallurgical microscope (Nikon, Mississauga, Ontario, Canada), equipped for differential interference contrast and dark field for epi-illumination.

Fibres retained on the surfaces of airways and alveoli: SEM analysis. The analysis by SEM allowed the investigation of fibres on airway and alveolar surfaces. Single fibres and small aggregates were found. Larger aggregates were observed at bifurcations. In the intrapulmonary conducting airways, 89% of the fibres were found to be completely covered by surface-lining layer material, and 11% were only partially covered. The average length and diametre of the fibres in the airways were 37.3 [+ or -] 19.4 [micro]m (mean [+ or -] SD; range, 7-104 [micro]m) and 1.1 [+ or -] 0.4 [micro]m (range, 0.3-2 [micro]m), respectively. The average length-to-diametre ratio was 34.0 [+ or -] 19.0 (range, 7-122). In the alveoli, 32% of the fibres were completely covered, 26% were partially, and 42% were not covered by the lung-lining layer. Some fibres appeared to project into the alveolar airspace or to cross it, and they touched the alveolar surface only with their ends. The average length and diametre of the fibres in the alveoli were 20.9 [+ or -] 8.3 [micro]m (range, 9-35 [micro]m) and 0.9 [+ or -] 0.3 [micro]m (range, 0.5-1.5 [micro]m), respectively. The average length-to-diametre ratio was 27.8 [+ or -] 14.0 (range, 10-52).

Fibre Retention in Lungs by TEM

The analysis by TEM allowed the investigation of fibres within the entire lung tissue. Because with this method we analysed two-dimensional transects of the fibres in tissue sections, we could observe the traces of cut cylinders, which were in all cases either circles or ellipses. A vitreous glass profile with irregular complicated shape can only originate from a broken fibre. Furthermore, if the profile's length-to-diametre ratio is also < 3, it is most likely a profile of a particle. From the 144 embedded tissue blocks, 96 contained intrapulmonary conducting airways as well as lung parenchyma, and these were used for analysis by TEM. A total of 335 MMVF 10a profiles, which are electron-dense and easy to recognise, were analysed on micrographs of ultrathin sections. We found 35 mostly fibre profiles in the conducting airways, and we registered 69 profiles in the alveoli, about half of them originating from fibres and the other half from particles. Eight fibres and/or particles were found to be phagocytosed by airway macrophages, and nine by alveolar macrophages. We observed 228 mostly particle profiles in the lumen of alveolar blood capillaries, and 1 within a pulmonary artery. One fibre profile was found within a type I epithelial cell, and the location of another one could not be identified. Neither fibre nor particle profiles were found in the interstitium.

In airways, 96% of the profiles were submersed in the surface-lining layer, whereas 4% appeared not to be covered by any material. In alveoli, 33% of the profiles were found to be completely, 13% partially, and 54% not covered by any surface-lining layer material. The profiles found in the lumina of blood vessels had blank surfaces.

Displacement of Fibres into an Aqueous Substrate by a Surfactant Film

The experiments in the Langmuir-Wilhelmy surface balance demonstrated that the extent of fibre immersion into the aqueous phase depends on the surface tension of the film at the air-liquid interface. At 30 mJ/[m^2], < 10% of the fibres were submersed into the aqueous substrate, whereas at surface tensions < 20 mJ/[m^2], almost all fibres (> 95%) were completely submersed. However, the displacement of fibres was found to depend also on size. Fibres shorter than 10 [micro]m were immersed into the aqueous subphase by approximately 50% at 25-30 mJ/[m^2] and were completely submersed at [less than or equal to] 20 mJ/[m^2]. Fibres with a length of 20-30 [micro]m were immersed to 50% at a surface tension of 15 mJ/[m^2] and to 95% at [less than or equal to] 10 mJ/[m^2].

Surface free Energy of Glass Fibres

The surface free energy of clean glass is several hundred millijoules per square metre. However, glass surfaces adsorb impurities (materials) suspended in the ambient air, which considerably reduces the surface free energy value of glass. Measurements on surfaces of glass slides have demonstrated that clean glass, after exposure for 15-20 hr to laboratory air, has a surface free energy dose to that of polystyrene (i.e., ~ 35 mJ/[m^2]). We therefore expect that the glass fibres investigated in the anatomy laboratory in Bern had surface free energies similar to that of polystyrene.

The microscopic analyses of inhaled and deposited MMVF 10a in hamster lungs described here show that in the conducting airways almost all fibres were submersed in the aqueous lining layer, whereas in the alveoli only about one-third of the fibres were completely covered by lining layer material. It has been shown previously that the displacement of polystyrene microspheres depends on the surface tension of the surfactant film at the air-liquid interface. The lower the film surface tension, the greater the extent of particle immersion into the aqueous subphase. The extent of immersion as assessed in the Langmuir-Wilhelmy surface balance demonstrated that immersion of MMVF 10a into the subphase by approximately 50% occurred for fibres shorter than 10 [micro]m at film surface tensions of 25 mJ/[m^2] and for those 20-30 [micro]m in length at 15 mJ/[m^2]. For submersion of the longer fibres, the film surface tension had to be [less than or equal to] 10 mJ/[m^2].

The calculation of the forces that are exerted on cylindrical MMVF 10a after their deposition on a liquid surface is somewhat more difficult than that for forces exerted on microspheres, because of the complex wetting bahaviour. First, fibres have infinite possibilities with respect to their orientation of hitting a liquid surface. Second, MMVF 10a have variable dimensions and are straight or bent cylinders, meaning that forces exerted on them would have to be calculated for each single fibre. Third, the surface chemistry of the glass fibres does not seem to be homogeneous, in that the three-phase line (the line where the vapour, fluid, and solid phases join) appears to be corrugated. This can be observed by light microscopy, especially on long fibres; dry spots (not wetted by the surfactant film) are followed by partially immersed parts, and so on. From the model experiments in the Langmuir-Wilhelmy surface balance, it follows that glass fibres (MMVF 10a) are displaced into the aqueous lung-lining layer only below a film surface tension of 20 mJ/[m^2]. Assuming that clean glass has a surface free energy of several hundred millijoules per square metre, one would expect immediate wetting and submersion of glass fibres caused by a surfactant film of a surface free energy of approximately 25 mJ/[m^2]. However, glass surfaces tend to adsorb impurities (materials) suspended in the ambient air, such that the surface free energy of the fibres used in this study is expected to be considerably lower. As a consequence, we observed that such fibres are wetted and displaced only if the film surface tension falls below approximately 30 mJ/[m^2], which is consistent with a reduced surface free energy of the glass due to adsorption of impurities.

According to microscopic analysis, about 90% of the glass fibres were submersed in the surface-lining layer in the intrapulmonary conducting airways. In this lung compartment, fibre length seemed not to be a limiting factor for their displacement, because fibres up to 104 lam in length were found to be submersed. The in vitro experiments demonstrated that in order to promote total immersion of glass fibres of a length of 20-30 [micro]m, the film surface tension has to fall to approximately 10 mJ/[m^2]. This value is considerably below the surface tension of 32 mJ/[m^2] measured in the trachea of horses and substantially below the equilibrium surface tension of approximately 25 mJ/[m^2] for pulmonary surfactant films. Because surface tensions < 22-25 mJ/[m^2] can only be achieved by mechanically compressed films, the conclusion is that the surfactant film reaches at least temporarily a compressed state in the intrapulmonary conducting airways of hamsters. Because the alveolar surfactant film reaches minimum surface tensions below approximately 2 mJ/[m^2] under in vivo conditions, it is clear that there must be a surface tension gradient between the compressed alveolar surfactant film and that in the trachea or large bronchi, but the definition of the gradient remains unclear.

Direct surface tension measurements in the small airway compartments have not been possible so far with current techniques because these airways are not accessible. However, indirect evidence for surface tensions far below the equilibrium value of approximately 25 mN/m in the central airways of rabbits has recently been obtained by Bachofen and Schurch (2001), who demonstrated that the shape of bronchiolar macrophages beneath the surfactant film in the surface-lining layer depends on the local surface tension. The shape of the macrophages, determined by TEM, may serve in turn to estimate the airway surface tension at that location.

Because the surface tension in the alveoli reaches values < 2 mJ/[m^2] on expiration, one would expect all of the fibres in the alveoli to be completely submersed in the surface-lining layer. However, in the present study, we found that about one-third of the fibres in alveoli were not covered by the surface-lining layer. This may be due to geometric limitations. The geometry of alveoli in hamsters resembles spherical cavities with an average diametre of about 80 [micro]m. Long fibres may touch the inner surface with the tips of both ends and bridge the alveolar airspace. Such configurations have been observed. The fibres are immersed into the lining layer only at their ends. Therefore, partial coverage of fibres by lung lining material as observed by SEM may be explained by size and geometry of the fibres and the airspaces.

Dissolution of the fibres becomes possible after their displacement into the aqueous lining layer. In addition, it is likely that surfactant components or proteins are adsorbed onto the surfaces of the fibres, which may alter their clearance pathway or their toxicologic potency. The displacement of fibres into the surface-lining layer brings them into close contact with resident macrophages, which comprise a major clearance pathway for particles deposited in the alveoli and in small airways. In addition, the displacement of fibres toward the epithelium may facilitate particle uptake by epithelial cells and eventually translocation to interstitial and endothelial sites, which may have consequences for causing lung disease.

In the present study, about two-thirds of all vitreous glass profiles were found in the lumen of alveolar blood capillaries. The majority of these profiles had irregular shapes and angular surfaces and most likely represents particles originating from broken fibres. The particulate fraction of MMVF 10a in the aerosol was 10%. Such particles are in a size range to reach the alveoli by inhalation and to be deposited there. This is demonstrated by the observation that many more particle profiles were found in the alveolar than in the airway compartment. Currently, it is not known how the MMVF 10a particles entered the capillary lumen. There are several mechanisms by which particles can be displaced from one tissue compartment to another one. First, particles < 1 [micro]m in diametre or those persisting on lung surfaces have been shown to be taken up more readily by epithelial cells or to be translocated even to interstitial and endothelial sites. We found only one vitreous glass profile within the interalveolar septum, within a type I epithelial cell. Therefore, clearance of MMVF 10a through the epithelial cells and further translocation via the interstitium and endothelial cells into the lumen of blood capillaries is unlikely. This is consistent with observations of Brody et al. (1981), who showed that for minute asbestos fibrils in rats, the above-mentioned sequence of particle transport had occurred only 4 or more days after exposure. In contrast, in the present study, < 1 hr elapsed between the onset of the inhalation until lung fixation. Second, mechanical translocation (e.g., due to the formation of a pneumothorax) is another mechanism by which intact or broken fibres may have entered the blood circulation.

Asbestos and glass fibres piercing the air-blood barrier have been demonstrated by others. In the present study, this might have happened upon pneumothorax formation, but we did not see ruptured tissue near to the location where we found fibres. Finally, particles may be driven through the cell membranes, for example, through those of the type I epithelial cells, by interfacial forces (adhesive interactions). We demonstrated in a recent study that particles < 1 [micro]m in diametre are able to penetrate the cell membrane of erythrocytes by nonspecific mechanisms.

These observations are consistent with those made by Rimai et al. (2000), who showed that spherical glass particles in the micrometre size range are driven into a polystyrene substrate by adhesive interactions. For example, glass particles 4 [micro]m in radius become engulfed by the substrate by approximately 90%, whereas particles 10 [micro]m in radius become substantially less engulfed. These investigators discussed the possibility of total engulfment of relatively hard and small particles by a softer substrate.

Nevertheless, if the particulate fraction of MMVF 10a is cleared by the blood circulation, organs beyond the lungs should be carefully investigated after fibre exposure. Marsh et al. (1990) reported a significant increase in mortality for nephritis and nephrosis in the 1985 follow-up study of workers exposed to man-made mineral fibres. Sali et al. (1999) suggested the possibility of an increased risk of nonmalignant renal disease with duration of employment in a nonneoplastic mortality study of European workers who produced rock or slag wool. Moreover, they demonstrated that mortality from ischemic heart disease was increased in workers exposed to rock or slag wool and in continuous-filament workers. Yet, Sjogren (2000) suggested that the increased mortality from ischemic heart disease originates from the high concentration of fibrinogen caused by the low-grade inflammation after fibre inhalation, and Chiazze et al. (1999) did not find any association between respirable fibres and nephritis or nephrosis in a case-control study on Caucasians who worked in glass wool production plants.

The observation that fibres were preferentially deposited at bifurcations in airways and alveoli has also been reported for inhalation studies using other fibre types.

In summary, the present study shows that glass fibres are displaced by a surfactant film and immersed into the aqueous lining layer in airways and alveoli, if the surface tension reaches [less than or equal to] 10 mJ/[m^2], provided that the fibres come into contact with the inner surface of the lung. The geometry and dimension of the airspace and the fibres are therefore important factors for fibre retention. We conclude that the sequence of wetting and immersion of inhaled and deposited MMVF 10a fibres into the aqueous lining layer by the surfactant film at the air-liquid interface is similar to that of spherical particles and is crucial for the retention of any kind of particles and fibres. Particles of identical shape, including asbestos or ceramic fibres, may have a surface free energy similar to that of vitreous fibres under ambient conditions. Presumably, they are likewise displaced into the aqueous lining layer after their deposition in the lungs. Therefore, the increased pathogenicity of certain fibres can most likely not be explained by different retention mechanisms. Adsorption of surfactant components or proteins upon displacement into the aqueous phase by surfactant is possible and needs further investigation because it may be a crucial factor for health effects. In addition, the importance of clearance of the particulate fraction via the blood stream needs further investigation.

Chapter 4

Carbon Nanotubes

The worldwide funding devoted to nanotechnology research and development by governments, industry, and venture capitalists was estimated to be around US$9.6 billion in 2005. A large portion of this spending is still being allocated to the development of nanoparticulate materials because of their many novel physical and chemical properties raising high expectations for a variety of applications. One of these new materials is carbon nanotubes (CNTs), which have commercial expectations in different manufacturing sectors.

However, epidemiologic studies of air pollution suggest that particulate matter has a strong association with cardiopulmonary diseases. Research has shown that nanoparticles may enter the human body more easily and be more biologically active because of their larger surface area per mass unit compared with that of larger particles. The prospective widespread use of engineered nanoparticles in consumer products may increase environmental, occupational, and public exposures dramatically. Consequently, different stakeholders have raised serious concerns regarding health effects of engineered nanoparticles. Recent review articles on the toxicity potential of nanoparticles conclude that the toxicity of nanoparticles depends on specific physiochemical and environmental factors. Thus, the toxic potential of each type of nanoparticle has to be evaluated individually.

Here we review the currently available literature on the potential risks of CNTs to human health and the environment. We also investigated the life cycle of CNTs, as release into different environmental compartments may occur at the production stages as well as at the product's use and disposal stages, which may directly or indirectly lead to human exposure. However, the published literature revealed many unanswered questions. Therefore, we also systematically interviewed seven leading world-class scientists and integrated their contemporary knowledge into this review. This assisted us in identifying questions and developing recommendations. The scientists interviewed were key authors or project leaders who have investigated and reported the potential impacts of CNTs on human health or environment. Through this combined approach we are able to present an updated and contemporary knowledge base for scientific discussion. In this review, we use the term "carbon nanotubes" when addressing the general aspects of the material, which includes singlewalled carbon nanotubes (SWCNTs) and multiwalled carbon nanotubes (MWCNTs). Here "multi" is defined as two or more walls. The terms "SWCNTs" and "MWCNTs" are used in a specified manner.

Exposure to Carbon Nanotube Material

Exposure in Occupational Settings

Procedures for the handling of CNTs can result in aerosol release of these materials into the surroundings. MWCNT aerosols generally have diametres between 20 nm to > 200 nm, lengths from 1,000 nm to > [10^6] nm, and different shapes (straight, partly rigid, bent, curled, and partly flexible) that may appear single or in clumps or ropes. Only one published study has investigated the potential for SWCNTs to become airborne. A laboratory study by Maynard et al. (2004) investigating the physical nature of the aerosol formed during mechanical agitation was complemented by a field study of SWCNT release during handling of unrefined SWCNTs. The authors found that sufficient agitation of unrefined SWCNT material can release fine particles into the air, but the concentrations generated while handling the material in the field were very low (< 53 [micro]g/[m^3]).

The laboratory study also revealed that different SWCNT production methods produced different types of aerosols. The laser ablation process generated a more compact aerosol that was difficult to break down into smaller particles, whereas the HiPCO (high-pressure carbon monoxide) process generated a more extended material that was easier to break down into smaller particles and appeared to lead to higher airborne concentrations. Maynard et al. (2004) also found glove deposits of SWCNTs during handling that were estimated at between 0.2 and 6 mg per hand and thus concluded that large SWCNT containing clumps had the propensity to become airborne and could remain so for long periods. This may cause dermal exposure and health risks even in less well-protected areas.

Maynard et al. (2004) noted that production volume was very small (research facility) and that workers took great care to reduce product loss during handling of the material. However, CNTs contain catalyst metals such as nickel, which is associated with increased risks of cancer in the nose region. Cleaning operations can also lead to emissions. The cleaning of the production chambers is performed usually using solvents or water, tissues, brushes, and sponges that are discarded after cleaning. This waste carries CNTs into the waste stream, thereby possibly becoming a source of release into the environment.

Exposure Through Environmental Media

Exposure through environmental media is highly relevant for several reasons: a) The widespread applications envisioned for CNTs may lead to substantial production volumes, and consequently to increased emissions into the environmental compartments air, groundwater, and soil. b) The physical and chemical processes in the environmental compartments may alter the properties of CNTs, for example, abiotic factors such as ultraviolet light may alter the coatings of CNTs as observed with fullerenes and quantum dots. Consequently, this may also change the bahaviour of CNTs in the environment and thus influence their environmental fate and impact. c) CNTs are possibly one of the least biodegradable man-made materials ever devised. They are totally insoluble in water in pristine form and are lipophilic by nature. It is generally known that biopersistent and lipophilic chemicals may accumulate along the food chain; therefore, such a scenario should also be evaluated with CNTs.

In aqueous environments, SWCNTs clump together to form aggregates in the micrometre range; these aggregates do not change in size distribution with increasing salinity or temperature. However, the aggregation differs with pH changes in water and postsynthesis treatment of the SWCNTs with, for example, acid or surfactants. Both these studies found that pristine nanotubes formed stable aggregates, whereas acid-treated nanotube suspension showed greater dispersion variability over time, yielding looser structures at large-length scales and more compact structures at small-length scales. The addition of a surfactant to CNTs resulted in a hydrophilic interface at the tip of the nanotubes that significantly enhanced nanotube dispersion. In laboratory assessments designed to assess the potential migration in natural porous media, SWCNTs have been shown to have mobility and deposition bahaviours different from those of other nanoparticles (Lecoanet and Wiesner 2004; Lecoanet et al. 2004). SWCNTs functionalised to facilitate dispersion in water displayed the highest mobility together with water-soluble fullerol, whereas colloidal [C_{60}], anatase titanium dioxide (Ti[O_2]), and ferroxane were among the least mobile of the nanomaterials evaluated.

The large surface area of CNTs may cause other molecules to adhere and potentially pick up pollutants and transport these throughout the environment. Several studies have investigated different carbon nanomaterials as superior sorbents of organic pollutants, metals, fluoride, and radionuclide 243-americium [243Am(III)]. Yang et al. (2006) found a high adsorption capacity of polycyclic aromatic hydrocarbons (PAHs) with different types of CNTs (MW and SW). This finding indicates a potential effect on the fate of PAHs upon their release into the environment. The adsorption capacity was found with the order SWCNTs > MWCNTs > [C_{60}] and seemed to be related to the surface area, micropore volume, and the volume ratios of mesopore to micropore.

Oberdorster et al. (2006), in a preliminary study, investigated the ingestion of SWCNTs by the suspension-feeding worm Caenorhabditis elegans. SWCNTs moved through the digestive tract and were not absorbed by the animal. However, even if SWCNTs did stay in the digestive tract, these materials could move up the food chain, as these worms and other organisms are consumed by benthivores. SWCNTs have also been shown to be bioavailable to aquatic organisms, as both water-solubilised (wrapped with a synthetic peptide) and unsolubilised SWCNTs were detected in the fecal material collected from the digestive tract in the exposed fish. For the water-solubilised SWCNT-exposed fish, clumps of SWCNTs were also found on the gill, but similar clumps were not visible in unsolubilised SWCNT-exposed fish. However, the fish mistook the unsolubilised SWCNTs (floating on top of the water) for food and ingested them. Furthermore, because pristine CNTs are lipophilic, there is concern that they might be taken up by microbial communities and roots and, consequently, accumulate in plant tissues.

Carbonaceous nanoparticles, including MWCNTs, can also be formed by natural processes and anthropogenic combustion processes. Although these MWCNTs can be prime suspects in the pathogenesis of cardiopulmonary diseases induced by fine particulate matter, there are physical differences between combustion-generated and manufactured MWCNTs. These MWCNT structures may therefore be less important when the impacts of engineered CNTs are being assessed, as the studies to date suggest that when the properties of

CNTs are altered by engineering, changes in the environmental fate of and human exposure to CNTs occur through the different environmental media.

In Vivo Studies-Pulmonary Toxicity

The first in vivo study of fullerene soot containing CNTs did not find any inflammation in the respiratory tract of guinea pigs or change of pulmonary function 4 weeks after exposure (Huczko et al. 2001), but it is possible that the method applied in this study (not examining lung pathology) did not reflect the actual toxicity of the material. The followup study with analysis 90 days postexposure of six different types of MWCNTs administered in individual doses (12.5 mg) to guinea pigs found small differences between the types of MWCNTs (Grubek-Jaworska et al. 2006; Huczko et al. 2005). However, for all types of MWCNTs there were multifocal granulomas observed around the materials, inflammatory reactions of terminal and respiratory bronchioles, and in some animals, mild fibrosis in alveolar septa.

SWCNT soot generated by the laser ablation method and intratracheally instilled in rats produced transient inflammation, cell injury effects, and a subsequent non-dose dependent series of multifocal granulomas. In comparison, equal doses of quartz produced sustained pulmonary inflammation, cytotoxicity, and fibrosis in a dose-dependent fashion. The authors questioned whether the failed observation of a dose-dependent relationship between SWCNT exposure and formation of the granulomas could be the result of the method of instillation and suggested that it should be clarified in an inhalation toxicity study.

In an intratracheal instillation study of mice, three types of SWCNTs were investigated (Lam et al. 2004) two types made by the HiPCO method and one by the electric arc method. The results from all three types showed that regardless of the amount of metal impurities, dose-dependent lung lesions were characterised chiefly by interstitial granulomas. The study also showed that both SWCNTs and ultrafine carbon black were taken up by alveolar macrophages, but that the effects of these materials were different. Macrophages containing carbon black were homogeneously distributed over the alveolar space, but macrophages containing SWCNTs clustered to form granulomas in centrilobular locations. In comparison, quartz induced mild to moderate pulmonary inflammation, which was considered less severe than that induced with SWCNTs.

The consequent agglomeration of CNTs into nanoropes makes it difficult to manipulate and administer these particles to experimental animals without forming nonrespirable particles that can lead to mechanical blockage of the airways, as observed in other studies. Therefore, Muller et al. (2005) compared the pulmonary toxicity of ground and unground MWCNTs in rats, using asbestos (Rhodesian chrysotile) and carbon black as references. The length of the individual MWCNT was reduced from 5.9 to 0.7 [micro]m as a result of the grinding procedure. The distribution in the lungs after intratracheal instillation was different for the two types. Agglomerates of intact MWCNTs remained in the largest airways, whereas ground MWCNTs were much better dispersed on the lung tissue surface. Regarding toxicity, the study found that after 60 days there were indications of a higher degree of pulmonary inflammation with ground MWCNTs than that found with intact MWCNT-treated animals. The induced effects were dose dependent [bronchoalveolar lavage fluid (BALF) lactate dehydrogenase (LDH) activity and protein content]. Histologic and biochemical analyses demonstrated a fibrotic reaction (granulomas) in the bronchi with unground MWCNTs and in the alveolar space or interstitial tissue with ground MWCNTs. In summary, the findings of Muller et al. (2005) indicate that asbestos-ground MWCNTs and unground MWCNTs induce at least partially similar effects [inflammation, fibrotic reaction, and increased tumour necrosis factor (TNF)-[α]production], whereas carbon black showed a transient inflammation with elevated TNF-[α]. The authors therefore concluded that these data point to a specific toxicity related to the unique properties of CNTs.

In a pharyngeal aspiration study of mice by Shvedova et al. (2005), the effects of HiPCO produced SWCNTs purified to a carbon content > 99% were investigated. The SWCNT aggregate depositions were correlated with granulomatous inflammation, whereas interstitial fibrosis with alveolar wall thickening was observed to be greater at 60 days than at 28 days postexposure in lung regions distant from the SWCNT aggregates. SWCNTs were compared with equal doses of nanoparticulated carbon black and fine crystalline silica, which did not induce granulomas and alveolar wall thickening and caused significantly less inflammation. Furthermore, the authors investigated whether exposure caused damage to pulmonary cells. This was confirmed by an increased number of alveolar type II cells (type II cells replicate after alveolar type I cell death). Exposure to SWCNTs also resulted in accumulation of an oxidative stress biomarker (4-hydroxynonenal) 1 day after exposure and also in a time-and dose-dependent depletion of a major antioxidant, glutathione, which was most severe 1 day postexposure. The rapid and dose-dependent fibrogenic response in regions of the lungs distant from SWCNT aggregates and in the absence of persistent inflammation was a unique finding in this study.

Initially, Shvedova et al. (2005) associated the interstitial fibrosis with deposition of dispersed SWCNT structures and pointed out that the mechanism for this response differs from the classic fibrogenic particle response in that it is not driven by chronic inflammation and chronic activation of alveolar macrophages. Additionally, the authors found that if pathogenic bacteria were inhaled by mice in combination with SWCNTs, bacterial clearance from the lungs was significantly slower. This indicates that SWCNTs, in addition to their primary effects, may also diminish general resistance to pathogenic attacks.

Dermal Toxicity

There is only one published study in vivo on the dermal toxicity of fullerene soot containing CNTs. Forty volunteers were subjected to a patch test, and four albino rabbits were subjected to an eye test. The study found no evidence of the induction of any response; thus, the authors concluded that soot containing CNTs is not associated with any risks. However, information is insufficient on CNT material characterisation and the study design for adequate validation of these results.

Translocation

There are some contradictory findings on the kinetics of water-soluble functionalised and radioactively rabelled CNTs in the body. Wang et al. (2004) showed that when hydroxylated SWCNTs with radioactive iodine-125 atoms were injected into mice, the SWCNTs behaved like small molecules, passed easily through a number of compartments, accumulated especially in bone, and were distributed throughout the whole mouse body except the brain. Eighty percent of the total dosing of the SWCNTs was excreted by the feces and urine after 11 days. In contrast, Singh et al. (2006) found no accumulation of chelated diethylentriaminepentaacetic and indium-111-rabelled SWCNTs or MWCNTs in mice after 24 hr. The biological bahaviour of functionalised (f)-CNTs is therefore comparable to small molecules. They are removed from the blood through the renal excretion route.

No toxic side effects or mortalities were observed, and the excreted f-SWCNTs and f-MWCNTs were intact. Both studies are valid only for functionalised tubes and cannot be interpolated for nonfunctionalised tubes because of their hydrophobic nature. The difference in the distribution in bone is probably because the functionalisation of the tubes was not the same, which influenced bahaviour in the organism.

Histopathologic Analysis ex Vivo

Sato et al. (2005) investigated the influence of different MWCNT lengths on the cytotoxicity of MWCNTs in the human THP-1 leukemia cell line in vitro and in subcutaneous tissue of rat in vivo. In a long-term assay in vivo, the authors observed 4 weeks after surgery that an increased inflammatory response was established only by the 825 nm MWCNTs, as indicated by the formation of granulation tissue. Most of the 220 nm MWCNTs were observed in phagocytes, with many of these recognisable in lysosomes. Conversely, the 825 nm MWCNTs were also observed in the intercellular space.

In Vitro Studies

The methodology of how the toxicity of CNTs is evaluated depends strongly on how CNTs are administered to the cells (i.e., homogenously dispersed or not, with or without surfactant, type and concentration of the surfactant). Furthermore, CNTs contain contaminants that may be bioactive per se. Often the exact methodology and CNT characteristics are not described, which makes it difficult to compare available data.

Pulmonary Cytotoxicity

In an in vivo study performed with nanoparticles such as Ti[O_2], Geiser et al. (2005) showed that these particles could be taken up by the lung, passed through the air-blood barrier, and translocated into the bloodstream. This pathway cannot be excluded for CNTs and may be supported at the single-cell level. SWCNTs suspended in Pluronic F108 (a nonionic surfactant) were incorporated by peritoneal macrophage-like cells using near-infrared fluorescence microscopy. Cherukuri et al. (2004) found that 1.5 [micro]g of 7.3 [micro]g/mL SWCNTs were taken up within 24 hr by macrophage cell cultures grown in 24-well culture dishes. The fastest reaction evoked by SWCNTs in macrophages was an oxidative stress that occurred within hours. The improved properties of nanomaterials also resulted in a catalytic activity. This catalytic activity may contribute to a new aggressive form of longterm toxicity. Oxidative stress is an imbalance between the production of reactive oxidising species (ROS) and their degradation by antioxidants. The intracellular equilibrium may be disturbed by the presence and/or uptake of nanomaterials. The concentration of ROS may be increased by the particle itself or by the disturbance of the ROS degradation pathway. Both cause an additional production of ROS, which interacts uncontrollably with the cell membrane, DNA, and/or other cell compounds, severely damaging these cell compounds.

According to Muller et al. (2005) and as described in previous sections, the adverse effects of MWCNTs depend on the length of the material used in vivo. Similar to their in vivo study, the authors found that long untreated tubes using Triton X-100 as the vehicle evoked an LDH release of nearly 20% at a concentration of 100 [micro]g MWCNTs/one 24-well of a 24-well plate, whereas the short ground MWCNTs induced a dose-dependent LDH release up to 35% greater than the corresponding vehicle (Triton X-100) treatment. This suggests that also in vitro the short MWCNTs are more toxic than the long ones. Interestingly, TNF-[α]was induced in vivo (measured in BALF) as well as in vitro (measured in preactivated peritoneal macrophages), indicating that TNF-[α]could be a reliable marker for future in vitro studies.

A comparative in vitro cytotoxicity study of several manufactured nanoparticles and nanotubes revealed that aggregated SWCNTs/ ropes, MWCNT raw material, and MWCNT aggregated tubes suspended in 5 [micro]g/mL dimethylsulfoxide (DMSO) affected cell viability in murine lung alveolar macrophages at nearly similar concentrations. Compared with different oxide particles of differing quality such as aluminium oxide, zirconium oxide, or ferric oxide, the SWCNTs and MWCNTs expressed comparable median effective concentration ([EC_{50}]) values (micrograms per milligrams), which is analogous to the same values obtained with asbestos used in the same studies as reference material. However, it is speculative to generalise these results in terms of health risks to humans, as the presented comparison study is based only on short-term effects during a 48 hr period. The physiologic relevance of these results remains to be determined.

In addition to the size, shape, and novel physical and chemical properties, the degree and type of agglomeration are important factors in the cytotoxicologic assessment of CNTs. Well-dispersed SWCNTs were less cytotoxic than micrometre-sized agglomerates of SWCNTs. To determine which carbonaceous nanomaterials have the most severe effect, comparison studies were conducted with the same biological model system. Jia et al. (2005) compared SWCNTs, MWCNTs, and [C_{60}] fullerenes suspended in cell culture medium and found that the cytotoxicity is apparently based on SWCNTs > MWCNT > quartz > [C_{60}]. Fiorito et al. (2006) in a second study on murine and human macrophages claimed that graphite had the most severe effect, followed by SWCNTs and [C_{60}] fullerenes.

Dermal Cytotoxicity

To emphasize the dermal toxicity of CNTs, Monteiro-Riviere et al. (2005) exposed human epidermal keratinocytes to MWCNTs that were produced without further purification processes. These particles were taken up by the keratinocytes in vacuoles where MWCNTs retained their structure, as demonstrated by transmission electron microscopy analysis. The cell viability parameters were reduced by 400 [micro]g/mL MWCNTs to as much as 20% after a 24-hr exposure. The expression of interleukin (IL)-8 was increased up to 6 times (400 [micro]g MWCNTs/mL) in a dose-dependent manner after that period compared with the corresponding control cell cultures.

A detailed characterisation of the molecular mechanism evoked by exposure of human skin fibroblast cells to 0.6 and 0.06 [micro]g MWCNTs/mL was evaluated by gene microarray analysis. This analysis showed that MWCNTs induced cell cycle arrest and increased apoptosis and necrosis. Of the genes evaluated, 216 genes changed their expression levels after treatment with MWCNTs at 0.6 [micro]g/mL. The most significant gene categories were Golgi vesicle transport, protein metabolism, secretory pathway, fatty acid biosynthesis [G_1]/S transition of mitotic cell cycle, and cell homeostasis. The fact that these cellular processes were affected by the presence of MWCNTs indicates that this stress has a remarkable influence on cell performance.

Shvedova et al. (2003) exposed human keratinocytes to HiPCO-produced SWCNT material containing 30% by weight of iron. One of the first reactions on SWCNT treatment, which occurred 18 hr after incubation, was an increased oxidative stress, as indicated by the presence of free radicals. The decrease of the total antioxidant reserve and the reduction of vitamin E as well as the increase of the lipid peroxidation products compared with the control cell culture support strongly the presence and the damage of the oxidative stress within the cell cultures. Shvedova et al. (2003) claimed that this oxidative stress was the reason for the reduced cell viability.

In contrast to the work presented previously in which CNTs were suspended by sonification in the presence of a solvent or cell culture medium, Sayes et al. (2006) functionalised purified SWCNTs before dispersing them in the cell culture medium. They produced SWCNT-phenyl-sulfate, SWCNT-phenyl-sodium sulfate ([SO_3]H), and six mixed SWCNT-carbon-sulfate and SWCNT-phenyl-[SO_3]H samples with ratios of 18, 41, and 80, respectively. According to the authors only water-dispersible SWCNTs were of considerable interest for biological applications. The suspensions obtained were used to treat human dermal fibroblast cultures. The pristine unmodified SWCNTs evoked a rate of 80% cell death at a concentration of 20 [micro]g/mL, whereas functionalised SWCNT-phenyl-[SO_3]H induced < 5% cell death at a concentration of 2,000 [micro]g/mL. Cytotoxicity was also strongly dependent on the type of sidewall groups: SWCNT-phenyl-[SO_3]H was less toxic than the SWCNT with phenyl-(COOH)2 group. The general conclusion reached by Sayes et al. (2006) was that an inverse relationship exists between the toxic potential and the degree of side wall functionalisation. This is an important issue for further applications of CNTs in nanomedicine. The intensive skin cell-CNT interaction studies emphasize that penetration of CNTs is still not understood.

In Vitro Studies with Neurons

Polym-aminobenzene sulfonic acid and ethylenediamine functionalised MWCNTs (MWCNT- PABS; MWCNT-EN) coatings were used to coat glass cover slips as substrates for neuronal growth. To achieve this coating, the tubes were suspended in water and glass cover slips were coated with the suspension. After the water was evapourated, hippocampal neurons were seeded on the cover slips. After 3 days of incubation the number of branches per neuron, total neurite length, and the number of neurites per neuron of the cultivated neurons were increased on functionalised MWCNTs but not on pristine MWCNTs. However, it must be noted that the performance of the neurons grown on polyethyleneimine (positive control) was better than that of neurons grown on MWCNT-coated surfaces. Additionally, Hu et al. (2004) claimed that the surface charge of the modified MWCNT can be used to control neurite outgrowth. An increase in neuronal outgrowth and branching was achieved by the functionalisation of SWCNTs with polyethyleneimine (SWCNT-PEI) compared with the modified MWCNTs. The authors described the morphology of these nerve cells in detail, but the functional proof, namely, in how far these cells exhibited spontaneous action potential and/or were able to be stimulated, is still lacking. All these studies reflected a positive effect of SWCNTs on neurons, but compared with other cytotoxicologic assessments, these tubes were immobilised on glass. A similar study was conducted by Liopo et al. (2006), who assessed growth and attachment of the neuroblastoma glioma NG108, a model neuronal cell, on unmodified SWCNT substrates and/or substrates from SWCNTs modified with 4-benzoic acid or 4-tertbutylphenyl functional groups using a simple functionalisation method. Liopo et al. (2006) found that SWCNT films support cell growth but at a reduced level compared with that of tissue culture-treated polystyrene.

Therefore, all studies mentioned here investigated in principle the degree to which nanostructured surfaces with chemical characteristics of CNTs affect the outgrowth of nerve cells and not the toxicity of suspended CNTs. In vitro cytotoxicity studies in different cell types. For human embryo kidney (HEK293) cells it was reported that in 0.5% DMSO, suspended SWCNTs are able to inhibit cell proliferation and to decrease cell adhesive ability in a dose and time-dependent manner. Additionally, these HEK293 cells exhibit an active response to SWCNTs such as the secretion of a 20-to 30-kDa protein with unknown function. After SWCNT treatment, further observations were made that were analogous to those in the reports in the previous sections. These included observations on cell cycle arrest in the [G_1] phase, up-or down-regulation of cell cycle-associated genes, and formation of apoptosis/necrosis. The degree to which the presence of the DMSO affects the cell surface and cell performance is unclear in the present study.

As described previously, the surface chemistry strongly affects the toxic potential of CNTs. For example, the cell viability of human T lymphocytes was decreased to 80% by exposure to 400 [micro]g/mL of oxidised MWCNTs (oxidised by refluxing the MWCNTs in concentrated nitric acid). The reduction in cell viability was explained by the increase of apoptotic cells after the treatment. In comparison pristine MWCNTs reduced viability by up to 40% and carbon black by only 15% at the same concentration. Interestingly, the oxidised MWCNTs appeared to be shorter and straighter.

Several studies were conducted in which SWCNTs were modified by adding various biological molecules such as phospholipids conjugated with polyethylene glycol (PEG)-folic acid, PEG-[NH_2], PEG-S-SDNA, etc. In all the studies the water dispersibility increased and the tubes were incorporated in HL60 and HeLa cells. Depending on the modification of the SWCNTs, these tubes could act as multifunctional biological transporters, for example, DNA or siRNA. Unfortunately, no quantitative measurement was performed to allow a statement and comparison of these modified SWCNTs regarding the uptake rate and induced toxicity.

Carbon nanotubes have also been proposed by Price et al. (2004) as a possible new orthopedic/ dental implant surface material because of their unique mechanical, electrical, and cytocompatibility properties. Cell viability and number of human osteoblast CRL-11372 were determined after 3, 6, 11, and 24 hr of incubation by ethidium homodimer and calcein AM staining. The authors used conventional fibres (diametre > 100 nm) and nanoscaled fibres (diametre < 100 nm). The fibres agglomerated within 1 week in cell culture media to ropes of about 340 nm in diametre in the case of conventional fibres and about 670 nm in the case of nanoscaled fibres. The nanoscaled carbon fibres appeared to influence osteoblast viability less than their conventional dimensioned counterparts. The suspended carbon fibre agglomerates were taken up by the osteoblasts and were incorporated in lysosomal vacuoles (Price et al. 2004). This future application of carbon nanotubes as implant material is in itself a promising issue, but one of the key questions is do carbon nanotubes influence the differentiation of osteoblast progenitor cells and to what degree is the formation and activation of osteoblasts affected?

In conclusion, CNTs were taken up by different cell types and evoked diverse effects in the cells. A first and fast reaction is the formation of free radicals (oxidative stress), which has been suggested as a key factor in further cell reactions. Several cell biological effects and changes in gene expression patterns were described, but because of the different CNT material and suspension procedures used, quantitative and comparison statements on the cytotoxicity of CNTs in different cell types and tissue are nearly impossible. Focused studies on inhalation exposures are needed to evaluate in vitro studies on their reliability and to build up the epidemiologic databases.

In the present review, we have investigated the state of knowledge regarding the impact of CNTs on human health and the environment.

At the time of the review, there were approximately 50 studies focusing on the effects of CNTs on human health and environment, with the majority of them in vitro studies. Environmental impacts in particular have been investigated poorly. One of the findings of the expert interviews was that the current impact assessments have been conducted without any overall research strategy. Acknowledging this, the experts have called for a better coordination of research, which could provide interdisciplinary and complementary results.

There are different types of CNTs produced in and applied to products with a variety of physical and chemical properties and potential exposure routes. The knowledge base suggests that altering these properties induces different effects on environmental health. Additionally, these properties may change during the life cycle of a product containing CNTs, as external physical and chemical influences differ at each stage. Therefore CNTs may cause different environmental health effects depending on the life cycle of the product and fate of the product in the environment. Tracing the fate of CNTs from each stage of the CNT product life cycle may be an effective orientation for prioritizing research.

CNT Material

The particular properties of CNTs depend on the particular production process used. After synthesis the raw material contains nanoparticulate impurities that influence the toxicity of the species. Postsynthesis treatment alters various properties of CNTs such as length, purity, degree of aggregation, wall structure (doping), and surface functionalisation. These properties are thought to determine toxicologic-relevant factors such as particle size, mobility in the environment, chemical reactivity, persistence, and bioavailability. Hence, the impacts of engineered CNTs are likely to differ from those of naturally occurring CNTs and may additionally depend on the technical application and circumstances of release.

Exposure

Emissions of CNTs may, depending on the application, occur at all stages in the product life cycle: synthesis, production of intermediates, further processing, product use, and disposal. From the production sites, emission of CNTs into surrounding air depends on process control, handling procedures (yielding, bottling, packing), cleaning, safety installations and procedures (in case of leakages and accidents), and waste conditioning. These procedures may also lead to release of CNTs into the wastewater. Waste items such as contaminated hand gloves, packaging, or worn filter pads could disperse CNTs in cases of sufficient agitation. The CNTs released from the different stages may show very different characteristics such as length, surface properties, attachments, and agglomeration size. The variety of effects observed in the studies seems to result from the characteristics of the particular CNTs tested. It is therefore essential to scrutinize whether CNTs used in these studies are of the same type as the CNTs that organisms may be exposed to under real environmental conditions. A critical investigation of the types of exposures to CNTs before using CNTs in studies seems crucial for the information value of the study. Therefore, focusing on characterised exposure scenarios could be very useful when conducting toxicologic studies.

Environmental Compartments

The fate of CNTs in environmental compartments may differ depending on their specific properties such as surface chemistry, electrical properties, and oxidative potential. Studies show that functionalisation and state of aggregation are likely to influence the bahaviour in water and soil. It would therefore be useful to characterise the CNT type and form that could be released into the environmental compartments to assess the fate of CNTs. However, the physical and chemical conditions of the different environmental compartments (such as redox potential, pH, temperature, ultraviolet light, or synergistic effects with toxins) are also likely to alter the properties of CNTs or their functional attachments, and thus their environmental fate. Other molecules or particles stick easily to the surface of CNTs and may alter the the bahaviour of CNTs. Consequently, pollutants may be transported in the environmental compartments. Therefore, the question arises as to whether and to what extent it would be possible to foresee and control the effects caused by different CNT material properties in the environment.

Environmental Health end Points

Once CNTs are released into environmental compartments, preliminary studies have shown CNTs to be bioavailable to different organisms in addition to being biopersistent. One can therefore not exclude the possibility that CNTs may accumulate along the food chain. These studies emphasize a further need to study ecotoxicity as well as longer time-scale impacts covering bioaccumulation, biopersistency, and negative effects on reproduction. Once taken up by humans or other species, CNTs may cause oxidative stress, inflammation, cell damage, adverse effects on cell performance, and, in a long-term perspective, pathological effects like granulomas, fibrosis, and wall thickening. These effects have been observed time and dose dependently in the majority of toxicology studies.

However, a remarkable range of results has been observed in toxicology studies. Comparing these different studies is difficult because different CNT species and cell model systems have been employed. An indepth characterisation of the CNTs employed is crucial for interpretation of cytotoxicologic and toxicologic assessments. The contradictions in some of the results obtained could also originate from the types and degrees of agglomeration of the CNTs. In addition, the surfactants or surface coating may have a significant influence on the CNT-cell interaction, an effect that remained unaccounted for in several studies.

Strategies to enhance the explanatory power of future toxicology studies should comprise some standardisation of the CNT material, test procedures, and cell models/organisms employed. To that end, experts recommended using SWCNTs from arc discharge synthesis, as these SWCNTs show relatively good purity and structural uniformity. In addition, a set of toxic equivalency factors (TEFs) would be useful for comparing the toxicologic effects of different types of CNTs. TEFs have been developed to compare the toxicity of substance classes such as PAHs, which comprise a wide range of compounds. The establishment of TEFs for CNTs would need to consider nanospecific properties of the material (e.g., particle size, aspect ratio, agglomeration state) and would require investigation of mechanisms and dose-response dependency on the cellular level.

To what extent CNTs can enter the bloodstream, for example, through the alveolar passage, and whether organisms are able to eliminate pristine CNTs remain unexplained. To date, no biodegradation mechanisms for CNTs that support elimination from the organism have been investigated. However, agglomeration and immobilisation of CNTs within tissue have been observed, for example, in lung tissue. Some studies have shown that functionalisation may, in addition to influencing CNT mobility, also influence the degree of toxicity. Functionalisation may therefore be a key parameter for controlling the impact of CNTs on human health and the environment. The important considerations discussed above combine different scientific disciplines ranging from materials science to biology.

Integrating nanotoxicology with a life-cycle perspective will therefore be a prerequisite for the development of nanotechnology-based applications in a safe and responsible manner.

Health Effects of Combustion By-products

Incomplete combustion processes can emit organic pollutants, metals, and fine particles. Combustion by-products represent global human and environmental health challenges that are relevant not only in heavily industrialised nations, but also developing nations where up to 90% of rural households rely on unprocessed biomass fules for cooking, warmth, and light. These issues were address at the Seventh International Congress on Combustion By-Products, convened 4-6 June 2001 in Research Triangle Park, North Carolina. This congress included a diverse group of multidisciplinary researchers and practitioners who discussed recent develop-merits and future goals in the control of combustion by-products and their effects of exposure on human and ecological health. Participants recommended that interdisciplinary, coordinated research efforts should be focused to capitalise on the important potential synergisms between efforts to reduce the adverse human health effects linked to exposures to combustion by-products and broader efforts to reduce greenhouse gas emissions and save energy through efficiency. In this chapter we summarise the principal findings and recommendations for research focus and direction. Key words: combustion by-products, fine particles, genetic susceptibility, lung pathobiology, metals, Polychlorinateddibenzodioxins, polychlorinated dibenzofurans, public health.

Inefficient combustion processes result in incomplete combustion and can emit complex mixtures of organic pollutants, metals, and fine particles due to incomplete combustion. Although this concept generally brings to mind the image of a tall smokestack spewing a dark plume, combustion sources are greatly varied and quite often not related to industrial facilities. Combustion processes include manufacturing processes, residential fuel use for heating and cooking, and open burning, as well as the diverse internal combustion engines found in jets, construction equipment, cars and trucks, lawnmowers, and so on. Not only are combustion processes diverse, but the by-products emitted (defined as products other than the most thermodynamically stable reaction products, i.e., carbon dioxide and water) vary with process efficiency and fuel type.

Indoor and outdoor environments are widely contaminated by complex mixtures of combustion-derived gases and particles, and these pollutants create global human and environmental health problems. Because combustion sources are found worldwide, it is not only the residents of industrial cities in developed nations that face the environmental health risks from exposure to combustion by-products. Moreover, by-products of combustion processes have potential for wide-ranging impacts, not only on human health but on entire ecosystems and the global climate. These issues were addressed at the Seventh International Congress on Combustion By-Products. This congress included a diverse group of multidisciplinary researchers and practitioners who discussed recent developments and future goals in the control of combustion by-products and their effects of exposure on human and ecologic health. In this chapter we summarise the principal findings and recommendations for research focus and direction. A key need is the development of technologies to provide safe, clean, and efficient energy systems, with the ultimate goal of translating discoveries to public health intervention strategies.

Combustion By-Products: Impacts on a Global Scale

Environmental health policy has primarily focused on issues that are considered global in scale. Current policies emphasize environmental issues that are global in the sense that the environmental impact is the result of worldwide emissions or atmospheric formation of a single pollutant or small identifiable group of pollutants. For example, combustion-generated hydrocarbons and nitrogen oxides react in the atmosphere in the presence of heat and sunlight to form harmful concentrations of ground-level ozone. But it is not only cities with high numbers of industrial facilities or cars that are subject to high ozone levels, as winds carry nitrogen oxides ($N[O_x]$) emissions hundreds of miles away from their original sources.

Global Climate Change

With regard to global warming, the surface temperature of Earth has increased approximately 1.2°C since 1850, with 0.5°C of the change occurring since 1978. Predictions for future changes in our climate vary, but most scenarios predict a steady increase in global temperature throughout the twenty-first century. The magnitude of the impacts of global warming on human health, environmental health, and national economies remains unknown. Climate change will likely cause more economic damage in developed countries and more human health damage in less developed countries. There is a growing consensus among many scientists that global warming is, at least in part, the result of increasing levels of anthropogenic greenhouse gases [carbon dioxide (CO_2), nitrous oxide ([N_2]O), methane (C[H_4]), chlorofluorocarbon (CFC)] and tropospheric aerosols. Industrial, energy-producing, and residential combustion processes contribute significantly to global levels of greenhouse gases (GHG). Burning of fossil fuels is the major source of CO_2 emissions, as well as a major source of aerosols, especially black carbon, organic aerosols, and sulfates. Combustion processes are also major emitters of N[O_x] and sulfur oxides (S[O_x]).

Human Health Effects

The effects of combustion by-product pollution on human health are real and need to be addressed by sound scientific research. The practical benefits of efforts to reduce pollutant emissions from combustion processes would be realised immediately. Environmental health researchers have estimated that reducing emissions from older coal-fired power plants in the United States could provide substantial benefits to public health, including the avoidance of 18,700 deaths, 3 million lost work days, and 16 million restricted-activity days each year. By reducing emissions from nine older coal plants in the Midwest alone, it has been estimated that roughly 300 deaths, 2,000 respiratory and cardiac hospital admissions, 10,000 asthma attacks, and 400,00.0 person-days of respiratory symptoms could be avoided each year. Some scientists have also predicted that reducing GHG emissions in just four major cities could save 64,000 lives over the next 20 years (10-12). It has been suggested that investments in modern technology and air quality control in India and China could lead to reductions in tropospheric ozone and black carbon that would not only improve local health and agricultural productivity but also benefit air quality and the global climate.

Numerous studies in urban areas have documented the relationship between urban atmospheric particulate matter pollution emitted by industrial/transportation combustion processes and respiratory and heart diseases (13-16). Less attention has been paid to the significant risk of exposure to harmful combustion by-products resulting from the indoor use of solid fuel for cooking and heating. Simple stoves burning solid fuel (mostly biomass fuel) are used by about half of the world's people. Poor ventilation and inefficient combustion result in significant daily exposure of women and young children to fine and ultrafine particulate matter, carbon monoxide,

(CO), $N[O_x]$, $S[O_x]$, a range of volatile organic compounds including formaldehyde, and polycyclic organic matter. The World Bank has estimated that indoor air pollution is responsible for almost 50% of the burden of total disease resulting from poor household environments in developing countries.

Individual Susceptibility and Predisposition

Environmental health researchers have determined that most human disease results from a combination of environmental exposures and genetic variation. However, it is not fully understood why certain people develop disease when challenged with environmental agents and others remain healthy. Thus, a more complete understanding of how genetic characteristics affect the human response to environmental exposures is needed to improve approaches to the prevention and control of environmentally induced diseases. Research addressing individual susceptibility, especially with respect to diseases resulting from exposure to combustion by-products, is progressing on a number of fronts, from molecular biology to epidemiology.

Genetic variations may impact innate immunity the ability to launch an immune response to toxins, microbes, and microbial products with no previous exposure. This may play a role in determining the susceptibility of an individual to an environmental exposure. For example, nearly everyone is exposed to at least low levels of environmental endotoxin. This biologically active lipopolysaccharide (LPS) is a component of the outer membrane of gram negative bacteria and produces airway inflammation when inhaled. LPS has been associated with the development and progression of asthma and other forms of airway disease.

However, not everyone exposed to LPS will develop airway disease. A broad range of stable, reproducible physiologic responses to inhaled LPS has been documented among healthy, nonasthmatic people. Researchers investigating the genetic basis for the physiologic response to LPS have gathered the first direct evidence that mutations, in a transmembrane receptor for LPS (TLR4) are associated with decreased responsiveness to inhaled LPS. These findings suggest that changes in the gene sequence alter the innate ability of the host to respond to environmental stress, as people with the TLR4 substitutions may be more resistant to localised forms of endotoxin-induced inflammation (e.g., asthma) but more susceptible to a systemic inflammatory response initiated or exacerbated by LPS (e.g., gram negative infections, sepsis, and adverse outcomes).

In home and/or occupational settings, most individuals are exposed to combustion-generated particles, but relatively few people with a compromised respiratory or cardiovascular system respond with increases in morbidity and mortality. As a result, researchers suggest that genetic factors play a major role in determining individual susceptibility to particulate matter (PM) exposure. Studies combining genome-wide scans with microarray analysis have provided comprehensive assessments of adverse responses to environmental stimuli. For example, research into individual susceptibility in mice to nickel-induced acute lethal lung injury revealed that response to this metallic constituent of particulate matter and cigarette smoke is controlled by a region on mouse chromosome 6 with modifier genes suggested on chromosomes 1, 2, and possibly 9 and 16 (27,28). These chromosomal regions are near regions that also have been identified as controlling the response to ozone, suggesting that some of the same genes may be involved in both responses. This type of information may help clarify the role polymorphic variation plays in the determination of individual response to environmental exposures. Such knowledge may lead to improvements in preventive medicine, successful treatment interventions, and more effective regulatory policies.

Epidemiologic studies provide insight to aspects of individual susceptibility not directly related to genetic variability. These studies identify factors such as age, sex, occupation, nutritional status, lifestyle, and coexisting health conditions. These factors must be considered both independently and as potential contributors to complex interactions. For example, boilermakers were studied to evaluate that role of occupational exposure to particulate matter in the development of pulmonary and respiratory diseases. Boilermakers are routinely exposed to boiler fossil fuel ash as well as particles and gases emitted during welding activities. Boiler ash composition varies with the type of fuel (oil, coal, natural gas, or trash) and may include fine particulate matter, polycyclic aromatic hydrocarbons (PAHs), and metals such as vanadium, nickel, iron, zinc, chromium, and arsenic. As dictated by maintenance and construction needs throughout the course of a year, a boilermaker may work at a variety of plants that burn multiple fuel types. The results of a 2 year longitudinal study conducted to establish if exposure of boilermakers to particulate matter is linked to loss of lung function suggest that working at gas, oil, and coal-fired plants is associated with adverse effects on lung function. A short-term prospective study to assess the relationship between continuously measured particles with an aerodynamic diametre (i.e., diametre of a sphere of unit density that has aerodynamic bahaviour identical to that of the particle of interest) < 2.5 [micro]m and cardiac autonomic function in a group of boilermakers exposed to fuel-oil ash and metal fumes revealed decreased variability in heart rate after occupational exposure, indicating autonomic dysfunction. This finding demonstrates the effects of fine particulate matter on the autonomic nervous system and may provide physiologic evidence for the mechanism linking particulate matter and increased morbidity and mortality.

Combustion By-Products: Concerns and Current Research

The physical state and chemical composition of combustion by-products vary with the type of combustion source, the efficiency of the combustion process, and the fuels that are burned. Combustion by-products can include CO, N[O_x], PAHs, polychlorinated dibenzo-p-dioxins (PCDDs) and dibenzofurans (PCDFs), particulate matter, and metals. Each of these pollutants has been associated with adverse impacts on human health, the environment, or the global climate (33-35). Researchers and regulators are paying particular attention to several combustion by-products, and a wide range of studies is being conducted to increase our understanding of the contaminant levels, risks of exposure, health impacts, and mechanisms of disease etiology.

Particulate Air Pollution

Particulate air pollution is characterised by both particle size and chemical composition. Fine particles [those with an aerodynamic diametre < 2.5 [micro]m (P[$M_{2.5}$])] are derived chiefly from combustion processes or are formed as secondary combustion particles including sulfates and nitrates. Particles in the approximate size range of 0.2-0.5 lam can be transported long distances and readily penetrated indoors, resulting in ubiquitous exposure. Fine particles are small enough to be breathed deeply into the lungs, exposure is widespread, difficult to avoid, and occurs throughout the entire life. The public health significance of fine particulate air pollution is substantial. One estimate of average population decrease in life span due to exposure to fine particulate air pollution for lifelong residents of some of the developed world's more polluted cites is 1-3 years.

Because no diagnostic test can prove causality, epidemiologic studies are used to evaluate the relationship between natural variability in exposure and variability in rates of illness. Many acute and chronic population-based epidemiologic studies have been conducted, and the overall evidence indicates a probable link between fine particulate air pollution and adverse effects on cardiopulmonary health, suggesting that all individuals who are chronically exposed may ultimately be affected. To investigate the physicochemical characteristics of particulate air pollution that effect the cardiorespiratory system and the mechanisms of action, epidemiologists examined the relationship between hospital emergency room data and ambient air monitoring data. Fine particulate matter was fractionated to study the impacts of sulfates, acidity, water-soluble metals, organic matter (OM), and elemental carbon (EC). The researchers observed a statistically significant, positive association between cardiac dysrhythmia and CO, coarse PM, and EC < 2.5 [micro]m (P[$M_{2.5}$] EC), as well as between all cardiovascular diseases and CO, P[$M_{2.5}$] EC and OM < 2.5 [micro]m (P[$M_{2.5}$] OM). Although covariation of many of the air quality indices may confound such analyses; the study provides one of the first assessments of PM components in relation to emergency room visits.

Relationships between exposure to fine particles and genetic mutations have just recently been established because epidemiologic studies require much longer monitoring periods and broader monitoring of possible end points. It has now been demonstrated that each elevation of exposure to 10 [micro]g/m3 of fine particulate air pollution is associated with an 8% increase in lung cancer mortality and a 6% increase in cardiopulmonary mortality. Thus, the fine particle-associated risk of death due to cancer is apparently greater than the risk due to cardiopulmonary disease. The U.S. Environmental Protection Agency's (EPA) current air standards for fine partite air pollution are based on the older study of cardiopulmonary disease and do not include the greater risk of death due to cancer.

Regulatory and industry efforts to control emissions of fine particulate pollutants depend on an accurate assessment of the relative contributions of different types of emission sources to ambient pollutant concentrations. This assessment is based on emission inventories compiled from source data and air quality models. Research has been conducted to improve methodologies for measuring primary emissions from a broad range of stationary and mobile sources and to evaluate post-emission gas to particulate conversions. Recent modifications of air quality models enable scientists to monitor the fate of particles emitted from different source types and calculate the contributions from each source to ambient air quality for use in the design of emission control programs. Advanced development of receptor-source models has complemented these source-receptor methods. Much of the critical work in the advancement of fine particulate matter sampling methods, development of air quality models, and development of defining source apportionment through "fingerprinting" was conducted by the late Glen R. Cass.

Ultrafine Particulate Air Pollution

Scientists have documented health effects of occupational exposure to ultrafine particles (those with an aerodynamic diametre < 0.1 [micro]m) , but the impacts of exposure to ultrafine particles in urban air from combustion sources such as automobiles, natural gas combustion, and electric engines remain unknown. Ultrafine panicles contribute little to the total particulate mass in urban air but are very high in number, reaching several hundred thousand particles per cubic centimetre. Ultrafine particles 20 nm in diametre have the highest deposition efficiency in the alveolar region of the lungs; ultrafine particles can rapidly penetrate the epithelium so that they can be detected in the interstitium of the lung shortly after exposure. Because ultrafine particles are always present in the urban atmosphere, it is likely that they play a role in causing acute lung injury in sensitive parts of the population. As a result of their small size, ultrafine particles have a large surface area and such particles may become coated or "enriched" with contaminants in combustion flue gas or the atmosphere including toxic metals (lead, cadmium, arsenic, chromium, and zinc), sulfur, PAHs, partially oxidised hydrocarbons including oxy-PAH (OPAH), and other heteroatom-containing species such as PCDDs and PCDFs.

Fine particles also stabilise biologically active, organic radicals that induce oxidative stress and DNA damage. The combination of PM-borne, persistent semiquinone-type radicals or organics and metals can lead to interactions resulting in the generation of reactive oxygen species (ROS) and other secondary, biologically active species. These findings represent a possible mechanism for the genesis of cancer due to exposure to fine particulate air pollution. Research with rodent models indicates that, per given mass, ultrafine particles administered to the lung cause a greater inflammatory response than larger particles and that coagulation of ultrafine particles to larger particles significantly decreases their toxicity. The pulmonary inflammatory potential of ultrafine carbonaceous particles can be significantly enhanced in rodents by gaseous co-pollutants such as ozone. In addition, surface chemistry may play an important role in ultrafine particle toxicity. Susceptibility to the inflammatory effects of ultrafine particles in rats and mice increases significantly with age and/or a compromised/ sensitized respiratory tract. It also appears that after deposition in the lung, ultrafine panicles are not efficiently phagocytised by alveolar macrophages. Translocation of ultrafine particles to the liver and of ultrafine dye particles along neurons has been documented, suggesting that organ systems other than the lungs may be impacted by ultrafine particles.

Elemental Speciation of Arsenic

Arsenic is emitted into the atmosphere by combustion processes such as coal-fired power generation plants, household use of coal fuel, burning of vegetation, and volcanic activity. Arsenic is released into the atmosphere primarily as arsenic oxide ($[As_2][O_3]$) and exists mainly adsorbed on particulate matter. These particles are dispersed by the wind and are returned to the earth by wet or dry deposition. In well-oxygenated water and sediments, nearly all arsenic is present in the thermodynamically stable pentavalent state (arsenate). The toxicity of arsenic species varies greatly. Inorganic arsenic compounds are highly toxic, with the anionic form, arsine ($AS[H_3]$), being more toxic than cationic arsenite ($[As^{3+}]$), and arsenate ($[As^{5+}]$).

Some organic arsenicals are nontoxic to humans. As a result of this variation, evaluation of total arsenic concentration is inadequate for estimating risk from arsenic present in food and drinking water supplies. Quantitative methods for assessing specific forms of arsenic are necessary for risk assessment studies and relative source contribution estimates. Identification and quantification of the specific forms of arsenic often require the combination of two complementary techniques: efficient separation of the species of interest and detection/ quantitation. Systems using liquid chromatography (LC) followed by inductively coupled mass spectrometry (ICP-MS) can be used for elemental speciation at sub-nanogeam and sub-picogram levels. Evaluation of the LC-ICP-MS system for potential interferences revealed that chloride can contribute minimally to $[As^{5+}]$ determinations and potential chromatographic interferences from iron could be minimised by the addition of a chelating agent such as EDTA and this speciation method should satisfy U.S. EPA validation requirements.

Metals in Particulate Manner

$P[M_{2.5}]$ particles are generally attributable to emissions derived from combustion sources and are often coated with contaminants from combustion flue gas, including toxic metals. Recent studies provide novel insights into the early events in activation of signalling cascades that lead to altered, cellular function after exposure to metals present in ambient air PM. Exposure to combustion-derived metal species such as As, V, and Zn is known to disregulate phosphoprotein metabolism, resulting in the modulation of intracellular signalling cascades and transcription factors. Current research indicates that these PM metals activate epidermal growth factor receptor (EGFR) and increase levels of guanosine triphosphate-bound Ras in human lung cells, and Ras activation is required for Zn-induced signalling. Although Zn-induced activation of Ras requires EGFR, it can still occur in cells expressing kinase-inactive EGFR or EGFR that is truncated or missing the regulatory autophosphorylation domain. In addition, phosphorylation of EGFR induced by Zn can be inhibited by a Src inhibitor and Src phosphorylation is markedly increased following Zn exposure. These data suggest that EGFR signalling by metal ions does not occur through a direct effect on the receptor, but is due to the activation of Src kinase. These findings show novel mechanisms of action by PM metals and may provide insight relevant to the health effects of PM exposure.

Mechanisms of Chromate Mutagenesis

Chromium is a widespread environmental contaminant and a known human carcinogen. In the environment, Cr exists primarily in the trivalent [Cr(III)] and hexavalent [Cr(VI)] forms. Compounds of Cr(VI), but not Cr(III), have been found to be mutagenic and carcinogenic. Chromate ($Cr[O_4^{2-}$), the major ionic form of Cr(VI), readily penetrates cell membranes. Researchers are using shuttle vector systems in yeast, mammalian cells, and lung tissue of transgenic mice to investigate the mechanisms of chromate mutagenesis. Chromate is reduced by glutathione, and mutagenesis is influenced by tissue levels of glutathione. Chromate reduction is accompanied by the generation of highly reactive Cr(V), a reactive intermediate, and reactive oxygen intermediates which can cause DNA damage including intra- and interchromosomal deletions and base substitutions.

These results support the hypothesis that Cr-induced mutagenesis is due to the generation of reactive oxygen intermediates during the reduction of Cr(VI) by glutathione. In studies conducted with a transgenic mouse system, researchers noted a small but significant increase in mutagenesis in kidney DNA in addition to Cr impacts on lung tissues.

Dioxins

PCDDs and PCDFs are ubiquitous environmental contaminants. Dioxins are potent animal toxicants, and the complex mixture of dioxins to which people are exposed is characterised by the U.S. EPA as a "likely human carcinogen". These compounds have never been produced intentionally nor used commercially in the United States, but are by-products of both anthropogenic and natural combustion processes. Historically, municipal and commercial waste incineration, pesticide manufacture and use, and pulp and paper processing have been believed to be the major sources of emissions. Voluntary industry actions, combined with regulatory controls, have resulted in dramatic reductions from these sources in the United States. Currently, uncontrolled burning of residential waste and accidental fires are thought to be among the largest sources of PCDD/PCDF emissions in the United States.

PCDD/PCDF formation in combustion processes involves a series of complex mechanisms, and specific aspects of these reactions have been the subjects of controversy. In general, PCDD/PCDF formation is a catalytic reaction of carbon with oxygen and chlorine, strongly dependent on the type and amount of catalyst, temperature, duration of combustion, and the type and amount of carbon. On-going research will provide a more complete understanding of dioxin formation to formulate strategies to reduce emissions. The influence of the polyvinylchloride (PVC) content of the waste stream on the formation of PCDDs/PCDFs during municipal solid waste (MSW) combustion has been debated over the last several years. Special interest groups (e.g., Greenpeace and Physicians for Social Responsibility) called for de-emphasis on pollution controls and new emphasis on control of the waste stream, replacing PVC with alternative, chlorine-free materials.

A recent study demonstrated that PCDD/PCDF formation is controlled by variation in combustion conditions, not PVC content or the amount of chlorine in the waste. Research on methods to reduce PCDD/PCDF emissions from small-scale commercial boilers for energy recovery and waste disposal indicates that if sufficient high-quality fuel is included in the waste stream, dioxin/furan emissions can meet the new European Union directives without the need for cyclone cleaning. The impacts of temperature and residence times on PCDD/PCDF formation have also been examined, and studies conducted in a pilot-scale solid fuel bed reactor revealed different formation pathways for dioxins and dibenzofurans. Chlorination of dibenzofurans occurs after longer dwell times and at lower flue-gas temperatures. Nearly all of the chlorine in flue gas is present as hydrogen chloride, and a catalyst is required for chlorination to occur, suggesting that the activity of the catalyst controls the formation rate of chlorinated furans. Additionally, these studies showed that when bromine is added to the fuel, more brominated than PCDFs are formed. A separate study to investigate the impact of increased bromine levels in a waste stream containing flame retardants found the majority of tetra-halogenated dioxins/dibenzofurans formed during combustion were dibromo-dichloro-compounds.

Advanced Technologies: Applicability to Research on Combustion By-Products

Questions concerning combustion by-products that remain unanswered include specific issues regarding formation, control, transport, fate, exposure, and human and environmental health impacts. Powerful new analytical methodologies, including microarray technology and nanotechnology, are providing novel approaches to investigate these issues. These same advances are also fuelling the growth of new scientific disciplines such as genomics, proteomics, toxicogenomics, and metabonomics. These new paths of investigation have the potential to lead to information, and new research strategies to elucidate toxicologic mechanisms which will provide the answers to some of the questions surrounding the environmental health impacts of combustion by-products.

Nanotechnology, the fabrication of devices with atomic or molecular scale precision (a nanometre is one-billionth of a metre), will allow scientists to build biosensors using semiconductor, organic, and hybrid materials. Potentially, nano-biosensors could provide rapid response time, real-time monitoring, and the capability to analyse biologic phenomena at the cellular level. They may reveal valuable basic information about intracellular responses to the toxicants in combustion by-products and delineate the origins of their adverse health effects. External biosensors may be capable of monitoring emissions of combustion by-products using compact solid state modules. In addition, nanotechnology is vital to the fabrication of DNA microarrays, which are powerful, high-throughput tools for simultaneously monitoring the expression of thousands of genes or even the entire genome. In a microarray, copies of DNA, each from a separate gene, are printed on "platforms" which can be glass slides, silicon chips, or nylon membranes.

After exposure to a chemical, gene expression is assessed by measuring the binding of mRNA to the microarray DNA. Generally the mRNA is tagged with fluorescent components, creating patterns of fluorescent dots that reveal which genes are expressed. Microarrays have a sensitivity and dynamic range similar to Northern blot analyses. Microarray slide-to-slide and spot-to-spot variability are less than animal-to-animal variability, and assay reproducibility is continually improving. The vast data derived from studies with microarrays are playing a critical role in the emergence of genomics, toxicogenomics, proteomics, and metabonomics. For example, microarray technologies have provided the tools needed to support the new scientific subdiscipline of toxicogenomics, which combines genomics and bioinformatics to identify and characterise the relationships between chemical exposure and changes in genome-wide gene expression patterns in target cells.

Microarrays will likely be used in the immediate future to better address a number of key toxicologic issues, including mode of action, dose-response relationships, chemical interactions, and the use of gene profiling for hazard identification in chemical mixtures and human exposure assessment. Genomics, tolicogenomics, proteomics, and metabonomics have the potential to change the classic risk assessment paradigm. Microarray technologies will allow scientists to move past the current focus on stepwise identification of single biomarkers of exposure to a specific chemical, and characterise gene profiles that identify multiple environmental stressors. This knowledge will allow us to sort out the subtleties of multiple and cumulative exposures as well as investigate the effects of timing, duration, and sequence of exposures.

Genomics, toxicogenomics, proteomics, and metabonomics will provide significant amounts of scientific data, but this information will be very technical and will seldom provide unambiguous answers to legal and ethical questions. These data are vulnerable to misinterpretation by scientists, regulators, the media, and the general public. It is critical that scientists and scientific institutions make a commitment to stress effective communication of research findings to policymakers and the general public. It is vital to ensure that information is presented in a fashion that is appropriate for the audience, especially when the information relates to risk. The general public is frequently presented with information that is incomplete and/or misleading. It is critical to solicit and respond to input from stakeholders and community members when designing research programs. This inclusive approach helps build social investment in research results and the public health interventions needed to protect human and environmental health.

In addition, DNA microarrays will be used to rapidly classify mechanism-based toxicants by characteristic expression profiles and also identify genes that are under similar regulatory control. It may be possible to use microarrays to examine the molecular background of differences in individual susceptibility to certain toxic chemical groups. Although an emerging consensus suggests that many of the complex and prevalent diseases that humans develop occur as a result of multiple biologically unique gene-gene and gene-environment interactions, this conceptual framework is limited. In fact, the development of disease in humans, environmental and otherwise, is far more complex. Environmental exposures affect those that are vulnerable temporally (age), spatially (geographically), and by unique circumstance (co-morbid disease, nutritional status, and genetics).

Even this paradigm fails to address the complex interaction of endogenous and exogenous risks that ultimately interact to cause disease. Although the recent advances in human and molecular genetics have provided an unparalleled opportunity to understand how genes and genetic changes interact with environmental stimuli to either preserve health or cause disease, without accounting for the temporal, spatial, and other unique components of an individuals microenvironment, our understanding of environmental health will remain incomplete. These concerns are particularly applicable to the health consequences of combustion byproducts. Future research efforts need to embrace the complexity involved in etiology and pathogenesis of disease rather simply reduce the model to an understandable, but minimally applicable, gene-exposure-response relationship.

Regulatory and Policy Issues Related to Particulate Matter

Particulate matter is a criteria pollutant, which means that national ambient air quality standards (NAAQS) have been developed to protect public health, public welfare, and the environment. These standards must be reviewed every 5 years. In 1997 the U.S. EPA published a final rule revising the NAAQS for PM. While retaining the P[M_{10}] standard levels, the U.S. EPA added new P[$M_{2.5}$] standards on the basis of epidemiologic studies that related ambient PM concentrations to various health effects, including premature mortality, exacerbation of asthma, and other respiratory tract diseases, and decreased lung function.

Fine particles behave entirely differently from coarse particles and are much more difficult and expensive to control. Faced with the enormous financial implications of the 1997 NAAQS for $P[M_{2.5}]$, the American Trucking Associations, the U.S. Chamber of Commerce, and other state and business groups filed a challenge to the U.S. EPA's action. The U.S. Court of Appeals for the District of Columbia Circuit struck down major elements of the U.S. EPA's two NAAQS standards, but the Supreme Court upheld U.S. EPA's authority under the Clean Air Act to set national air quality standards that protect the American public from harmful effects of air pollution. On 26 March 2002, the District of Columbia Circuit Court rejected the remaining claims that the U.S. EPA's decision was arbitrary and capricious and not supported by the evidence. The implementation guideline for the $P[M_{2.5}]$ standards calls for completion of the scientific review in 2002; designation of nonattainment areas by 2005; submission of state implementation plans by 2008; and state compliance by 2017.

An additional regulatory effort impacting particulate matter emitted as a by-product of combustion processes is the Regional Haze Rules. According to the National Academy of Sciences, the average visual range in most of the western United States, including national parks and wilderness areas, is 100-150 km (about 60-100 miles), or about one-half to two-thirds of the natural visual range that would exist in the absence of air pollution. In most of the East, including parklands, the average visual range is less than 30 km (about 20 miles), or about one-fifth of the natural visual range. The primary causes of regional haze are sulfates, organic matter, elemental carbon (soot), nitrates, and soil dust. The Regional Haze Rules are an attempt to tackle the combined visibility effects of numerous pollution sources through a regional, rather than a purely local-source, approach. These rules require all states to make "reasonable progress" in reducing any effect this pollution has on visibility conditions in class I areas and to prevent future impairment of visibility. This regional approach calls for the reduction of non-$C[O_2]$ GHGs and is predicted to reduce emissions of key pollutants including $N[O_x]$, $S[O_x]$, and CO. Such reductions will have a positive impact on our climate and, just as importantly, have an immediate impact on our quality of life.

Critical international efforts to minimise or eliminate emissions of combustion by-products are illustrated by the 2001 Stockholm

Convention on Persistent Organic Pollutants (POPs). This agreement calls for the elimination of production and use of all intentionally produced POPs and for continuing efforts to minimise emissions of unintentionally produced POPs with the goal of ultimate elimination. The convention requires the establishment and maintenance of source inventories to demonstrate continuing minimisation of each country's total emissions. The United Nations Environment Programme is focusing efforts to ensure that all countries compile inventories that are complete and comparable. Incineration will undoubtedly be the most widely used method for destruction of POPs.

Ethical Issues Related to Combustion By-Products

As governments attempt to formulate appropriate responses to environmental issues, they must understand and consider a myriad of potential legal, ethical, and social issues. They must acknowledge that air pollution is a global issue and that regional and transoceanic transport of air pollutants impacts neighbours, even in distant regions. Environmental issues related to combustion by-products must be considered from a global public health point of view. It has been argued that although the common perception is that the biggest environmental problems lie in the industrial world, from the perspective of actual ill health, environmental health impacts are most severe in less-developed nations. In addition, the largest burdens of environmentally related diseases (including respiratory infections related to indoor combustion of biomass fuels) in developing nations are borne by children.

Developed nations have a variety of approaches to address these issues in less-developed nations. Many of the environmental health problems in the developing world are tightly interrelated with poverty. Developed countries could address these issues by increasing efforts to enhance socioeconomic development, which should result in healthier household environments. Additionally, industrialised countries should eliminate and reduce the hazards caused by their own economic and social development and promote new affordable technologies for developing countries. Faced with the reality of finite resources, it is imperative that policymakers focus attention on developing strategies that will yield the greatest impact.

A series of potential legal, ethical, and social dilemmas evolves from research advances. This concept is illustrated by the possibilities presented by scientific advances in toxicogenomic research.

Toxicogenomics will enable scientists to identify genetic variations and their impacts on individual response to environmental exposures. Such information could serve as the foundation for the developing strategies to prevent and control environmentally induced diseases. Alternatively, this information contributes to a suite of ethical, legal, and social challenges. Discovery that polymorphism of a specific gene is linked to an environmental disease could lead to discrimination or stigmatization. Health and life insurance companies might discriminate against individuals who carry a gene linked to a disease. Employers might terminate or reassign workers on the basis of such information. Because some polymorphisms are more common in certain populations than in others, it is possible that a particular type of genetic susceptibility might be associated with particular ethnic or social groups. This may threaten the employment and insurance options available to entire groups of people. The identification of a polymorphism that places an individual at a higher risk of developing an environmental disease could result in the individual being stigmatized as carrying a "defective gene" or as already ill, even though an environmental trigger is required for the development of the associated disease and the individual may never develop the disease.

Not only will policymakers have to formulate regulations to prevent discrimination and protect individual privacy, they must confront the challenges of promulgating fair and appropriate regulations based on new information at increased levels of complexity. Regulators will need to determine exposure limits, deciding if limits should be designed to protect everyone at the levels required to protect the most susceptible populations. They will have to address the question of what constitutes an acceptable risk. To address each of these issues appropriately, regulators require in-depth, accurate, and accessible scientific information.

Research on toxic combustion by-products may be characterised by its complexity: the number of agents, their interactions, their sources and origins, and their environmental fates. Powerful advances in technology and models have been achieved that now allow the study of this difficult problem, a problem that has been previously considered too complex and onerous to be addressed in regulations for all but a few sources and a few pollutants. Recommendations for research focus and direction identified during the Seventh International Congress on Combustion By-Products are presented below.

Recommendations for research policy priorities are:

- Interdisciplinary, coordinated activities should be developed among the governmental agencies responsible for public health, worker health, environmental and industrial engineering, environmental monitoring, biomedical research, health policy, and respiratory medicine. This is needed to more effectively target research that will reduce risks to both human health and the environment.
- Dialog should be promoted among policy makers, researchers, nongovernmental agencies, and research sponsors to address regional and local exposure issues. Coordinated, multisite, multipollutant research efforts are needed to specify and quantify the health effects of exposures to combustion by-products, as they are not adequately represented by exposure to criteria pollutants such as ozone.
- New combustion research programs should be designed to capitalise on the important potential synergisms between on-going international programs including efforts to reduce greenhouse gas emissions and those to reduce health-damaging emissions from solid fuel stoves. Such programs need to integrate the efforts of policy makers and economists as well as engineers, scientists, and health researchers.
- Additional attention needs to be paid to improve development and implementation of measurement and monitoring techniques. Improved diagnostics could lead to smart systems, better understanding, and efficient control/prevention strategies for better use in public health policy decision-making.
- Policy makers should focus their attention on the development of strategies to address the complex issues surrounding environmental health, specifically from the standpoint of its ethical, legal, and social implications.

Recommendations for a research framework are:

- An initiative on the origin and mechanism of genotoxicity of fine particles should be undertaken in light of the recent epidemiologic evidence for the correlation of exposure to airborne fine particulate matter and increased risk of lung cancer, as well as the paucity of published research on the source of this correlation.

- Advanced methods are needed for characterisation of the bioavailability of particle-associated pollutants. Fine particles can act as effective delivery systems for toxic chemicals, but the bioavailability of the toxins depends on the nature of their interaction with the particles. Advanced methods for characterising the nature of association and binding of stable molecular as well as radical species with various types of particles are essential.
- Research is needed to increase our understanding of mechanisms of pollutant formation and stabilisation at the gas-particle interface. Research on the rate of formation of gaseous pollutants and precursors as well as development of mechanisms of gas-particle interactions to form pollutants is strongly indicated. This will require development of more detailed mechanistic models and methods for identifying gas-phase and surface intermediates as well as products.
- Improved computational methods for modelling gas and particle properties are needed. Detailed reaction kinetic mechanisms of pollutant formation and fate require accurate thermochemical and reaction rate parameters. Development of reliable ab initio methods of calculation of these properties are necessary as semi-empirical methods appear unreliable.
- Rapid microarray techniques should be applied for assessing potential health impacts of individual pollutants and mixtures. The varied and complex products formed in combustion systems can potentially induce a variety of health impacts alone or in combination in a manner that has previously defied characterisation. Application of the rapidly advancing field of microarray technology to screening for various biologic end points of individual pollutants and mixtures should be used to address this poorly understood but critical component of environmental health.
- Additional research should be conducted to increase understanding of the composition, sources, and health impacts of fine and ultrafine particles. Atmospheric studies should determine the composition of fine and ultrafine particles, including nano-particles; determine the ratios of elemental carbon to organic carbon in airborne particulate matter, identify the main sources for fine, ultrafine and nano-particles, and identify alterations of fine, ultrafine and nano-particles by time and by location.

- Research in toxicology should be advanced to determine the fate of fine and ultrafine particles during inhalation and after deposition (dosimetry); determine translocation rates to other extrapulmonary tissues and influence of particle chemistry on such translocation; evaluate specific cellular and molecular mechanisms via both animal and in vitro studies; develop and use compromised animal models; and to apply toxicogenomics and proteomics in mechanistic studies.
- Epidemiology studies should include prospective cohort studies of specific compromised people (e.g., cardiovascular diseases, respiratory including chronic obstructive pulmonary disease (COPD), asthma). Clinical studies with surrogate or concentrated ambient particles should include the elderly and subjects with cardiovascular disease, asthma, and COPD, as well as include measures for endothelial dysfunction and other vascular parameters in addition to cardiac and respiratory end points.
- Advanced techniques are needed for speciation of toxic metals. The oxidation state as well as chemical form of metals affects their toxicity, increasing the importance of improved methods for speciation of metals, especially those associated with particles and surfaces.
- Concerted and focused efforts should be made to fulfill the potential of biomarkers to significantly reduce the burden of exposure and disease and to protect individuals from the uncertainty of risk, specifically in the area of exposure and effects of combustion by-products.
- Efforts should be focused to develop reliable and sufficiently widespread environmental and health monitoring systems. This is required both to effectively target actions to reduce outdoor pollution and to evaluate their effectiveness.

Chapter 5

Development for Genome Sequencing

The purpose of this Request for Applications (RFA) is to solicit grant applications to develop novel technologies that will substantially reduce the cost of genomic DNA sequencing. Current technologies are able to produce the sequence of a mammalian-sized genome of the desired data quality for $10-50 million; the goal of this initiative is to reduce costs by at least two orders of magnitude. It is anticipated that emerging technologies are sufficiently advanced that, with additional investment, it may be possible to achieve proof of principle or even early stage commercialisation for genome-scale sequencing within five years. A parallel RFA HG-04-003 solicits grant applications to develop technologies to meet the longer-term goal of achieving four-orders of magnitude cost reduction in about ten years.

The ability to sequence complete genomes and the free dissemination of the sequence data have dramatically changed the nature of biological and biomedical research. Sequence and other genomic data have the potential to lead to remarkable improvement in many facets of human life and society, including the understanding, diagnosis, treatment and prevention of disease; advances in agriculture, environmental science and remediation; and the understanding of evolution and ecological systems.

The ability to sequence many genomes completely has been made possible by the enormous reduction of the cost of sequencing in the past two decades, from tens of dollars per base in the 1980s to a few cents per base today. However, even at current prices, the cost of sequencing a mammalian-sized genome is tens of millions of dollars and, accordingly, we must still be very selective when choosing new genomes to sequence. In particular, we remain very far away from being able to afford to use comprehensive genomic sequence information in individual health care. For this, and many other reasons, the rationale for achieving the ability to sequence entire genomes very inexpensively is very strong. There are many areas of high priority research to which genomic sequencing at dramatically reduced cost would make vital contributions. 1) Expanded comparative genomic analysis across species, which will yield great insights into the structure and function of the human genome and, consequently, the genetics of human health and disease. Studies to date that have been able to compare small regions of several genomes, and "draft" versions of full genomes, have clearly demonstrated the need for much more complete data sets. While some of the needed data will be obtained over the next two or three years using existing DNA sequencing technology, and while costs will continue their gradual decline, the cost of current approaches to sequence acquisition will continue to limit the amount of useful data that can be produced. 2) Studies of human genetic variation and the application of such information to individual health care, which will also require much cheaper sequencing technology. Today, genetic variation must be assessed by genotyping the relatively few known differences at a relatively small number of loci within the human population. A richer and better characterised catalogue of such variable sites is being generated to support more detailed and powerful analyses.

While these methods are, and will become even more, powerful and likely to provide a significant amount of important new information, they are nevertheless only a surrogate for determining the full, contiguous sequence of individual human genomes, and are not as informative as sequencing would be. For example, current genotyping methods are likely to miss rare differences between people at any particular location in the genome and have limited ability to determine long-range information (e.g., genomic rearrangements). Therefore, new methods based on complete genomic sequencing will be needed to use genomic information for individual health care in the most effective manner possible. 3) While the genomes of a few agriculturally important animals and plants have been sequenced, the most informative studies will require comparisons between different individuals, different domesticated breeds and several wild variants of each species. 4) Sequence analysis of microbial communities, many members of which cannot be cultured, would provide a rich source of medically and environmentally useful information. And accurate, rapid sequencing may also be the best approach to microbial monitoring of food and the environment, including rapid detection and mitigation of bioterrorism threats.

Given the broad utility and high importance of dramatically reducing DNA sequencing costs, NHGRI is launching two parallel technology development programs. The first, described in this RFA, has the objective of reducing the cost of producing a high quality sequence of a mammalian-sized genome by two orders of magnitude. The goal of the second program is the development of technology to sequence a genome for a cost that is reduced by four orders of magnitude. For both programs, the cost targets are defined in terms of a mammalian-sized genome, about 3 gigabases (Gb), with a target sequence quality equivalent to, or better than, that of the mouse assembly published in December 2002.

The ultimate goal of this program is to obtain technologies that can produce assembled sequence (i.e., de novo sequencing). However, an accompanying shorter-term goal is to obtain highly accurate sequence data at the single base level, i.e., without assembly information, that can be overlaid onto a reference sequence for the same organism (i.e., re-sequencing). This could be achieved, for example, with short reads that have no substantial information linking them to other reads. While the sequence product of this kind of technology would lack some important information, such as information about genomic rearrangements, it would nevertheless potentially be available more rapidly and produce data of great value for certain uses in studying disease etiology and pharmacogenomics, and for comparative genomics between closely-related organisms. Therefore, both programs' objectives include a balanced portfolio of projects developing both de novo and re-sequencing technologies.

State-of-the-art technology (i.e., fluorescence detection of dideoxynucleotide-terminated DNA extension reactions resolved by capillary array electrophoresis [CAE]) allows the determination of sequence "read" segments approximately 1000 nucleotides long. If all of the DNA in a 2-3 Gb genome were unique, it would be possible to determine the sequence of the entire genome by generating a sufficient number (millions) of randomly-overlapping thousand-base reads and align them by overlaps. However, the human and the majority of other interesting genomes contain a substantial amount of repetitive DNA (short [tens to thousands of nucleotides], nearly or completely identical sequences present in multiple [tens to thousands of] copies). To cope with the complexities of repetitive DNA elements and to assemble the thousand-base reads in the correct long-range order across the genome, current genomic sequencing methods involve a variety of additional strategies, such as the sequencing of both ends of cloned DNA fragments, use of libraries of cloned fragments of different lengths, incorporation of map information, achievement of substantial redundancy (multiple reads of each nucleotide from overlapping fragments) and application of sophisticated assembly algorithms to align and filter the read information.

The "gold standard" for genomic sequencing is 99.99% accuracy (not more than one error per 10,000 nucleotides) with essentially no gaps. At present, the final steps in achieving that very high sequence quality cannot be automated and require substantial hand-crafting. However, recent experience suggests that the majority of comparative sequence information can be obtained from automatically generated sequence assemblies that have been variously identified as "high-quality draft" or "comparative grade." Therefore, while the ultimate goal is sequencing technology that produces perfect accuracy, the goal of the current program is to develop technology for producing automatically generated sequence of at least the quality of the mouse draft genome sequence that was published in December 2002.

Emerging technologies, collectively characterised as sequencing-by-synthesis or sequencing-by-extension, may be able to achieve large numbers of sequence reads by extending very large numbers of different DNA templates simultaneously, but generally only for a few tens of bases as currently practiced. Even if it is possible to extend these reads to several hundred bases, it will still be necessary to link those reads to achieve long-range sequence contiguity. For some purposes, long-range sequence contiguity may not be required. For example, the resequencing of genomes (determination of the DNA sequence for many individuals of a species after a reference sequence for that species has been determined), such as might be used for medical diagnostic purposes, could be achieved by aligning individual reads on the reference sequence.

However, short reads, particularly ones with lower per-base quality, can be very difficult to align given the nature of repetitive DNA and of closely-related gene families in complex genomes. Also, chromosomal rearrangements may be difficult to detect without high quality sequence information bridging the breakpoints with enough sequence to know in which repeat the breakpoint lies. The determination of single nucleotide polymorphisms (SNPs) and their phase (for haplotypes) also requites contiguity of varying length. The ultimate goal and a high priority for the National Human Genome

Research Institute's (NHGRI) sequencing technology development efforts, as exemplified in these two RFAs, continues to be de novo, assembled sequence. However, because of the value of resequencing for many future purposes, these RFAs also solicit the development of very inexpensive technology for very high quality re-sequencing (without assembly).

Most investigators interested in reducing DNA sequencing costs anticipate that a few additional two-fold decreases in cost can yet be achieved with the current CAE-based technology, with a realistic lower limit of perhaps $5 million per mammalian-sized genome. However, it is likely that this efficiency will only be achieved in a few very large, well-capitalised, experienced, automated laboratories. To achieve the broadest benefit from DNA sequencing technology for biology and medicine, systems that are not only substantially more efficient but also more usable by the average research laboratory are needed. One set of current technology development efforts is aimed at increasing parallel sample processing while integrating the sample preparation and analysis steps on a single platform. Thus, in one approach, lithography is used to create a large number of microchannels on a single device and to integrate an efficient sample injector with each separation channel.

Chambers for on-chip DNA amplification, cycle sequencing reactions and sample dean-up have been also developed, and experiments to integrate these steps, an approach that effectively places much of the actual process and process control onto the device, are being conducted in several laboratories. Attendant improvements in separation polymers and in fluorescent dyes will facilitate these developments. As these approaches are based largely on the experience of currently successful high-throughput CAE-based methods, they have potential to produce cost savings in the range of several factors of two beyond the CAE-based system itself. They also have the potential to widen the user base for the technology, as the infrastructure and knowledge needed to conduct relatively high-throughput sequencing, or clinical diagnostic sequencing, would be substantially reduced and simplified.

Other approaches to improving sequencing technology involve methods that are independent of the Sanger dideoxynucleotide chain termination reaction or of electrophoretic separation of the termination products. Two methods that were proposed in the early days of the HGP involve the use of mass spectrometry and sequencing by hybridisation. Both methods have been pursued, with some limited success for sequencing, but substantial success for other types of DNA analysis. Both continue to hold additional potential utility for sequencing, although certain inherent limitations will need to be overcome.

More recently, additional methodologies have been investigated. These may be classified into two approaches. One is sequencing-by-extension, in which template DNA is elongated stepwise and each extension product is detected. Extension is generally achieved by the action of a polymerase that adds a deoxynucleotide, followed by detection of a fluorescent or chemiluminescent signal; the cycle is then repeated. Modifications of this approach rely on other types of enzymes and detection of hybridisation of rabelled oligonucleotides. To obtain sufficient throughput, the method is implemented at a high level of multiplexing, e.g., by arraying large numbers of sequencing extension reactions on a surface.

A key factor in this general approach is the manner in which the fluorescent signal is generated and the system requirements thus imposed. Depending on the specific approach, challenges of template extension methods include the synthesis of rabelled nucleotide analogues; identification of processive polymerases that can incorporate nucleotide analogs with high fidelity; discrimination of fluorescent nucleotides that have been incorporated into the growing chain from those present in the reaction mix; distinction of subsequent nucleotide additions from previous ones; accurate enumeration of homopolymer runs (multiple sequential occurrence of the same nucleotide); maintenance of synchrony among the multiple copies of DNA being extended to generate a detectable signal, or achievement of sensitivity that detects extension of individual DNA molecules; and development of fluidics, surface chemistry, and automation to build and run the system.

Current efforts to develop such methods have produced, at best, short sequence reads (less than or equal to 100 bases), so a continuing challenge is to extend read length and develop sequence assembly strategies. NHGRI anticipates that the state of the art for this approach is sufficiently advanced that, with additional investment, it may be possible to achieve proof of principle or even early commercialisation for genome-scale sequencing within five years. It is anticipated that the cost of genome sequencing with this technology could be reduced by two orders of magnitude from today's costs. It is important to note that sequencing by extension is one prototype for achieving these time and cost goals, but other technological approaches may also be viable. Developing technology with which to reduce the cost of genome sequencing by 100-fold is the subject of this RFA.

A second alternative to CAE sequencing seeks to read out the linear sequence of nucleotides without copying the DNA and without incorporating labels, relying instead on extraction of signal from the native DNA nucleotides themselves. The most familiar model for this approach, but almost certainly not the only way to achieve 10,000-fold reduction in sequencing costs, is nanopore sequencing, first introduced in the mid-1990s. Generally, this approach requires a sensor, perhaps comparable in size to the DNA molecule itself, that interacts sequentially with individual nucleotides in a DNA chain and distinguishes between them on the basis of chemical, physical or electrical properties.

Optimal implementation of such a method would analyse intact, native genomic DNA molecules isolated from biological, medical or environmental samples without amplification or modification, and would provide very long sequence reads (tens of thousands to millions of bases) rapidly and at sufficiently high redundancy to produce assembled sequence of high quality. NHGRI anticipates that the science and technology needed to reduce sequencing costs by four orders of magnitude, whether by the nanopore or some other approach, will require substantial basic research and development, and may take as long as ten years to achieve. Reaching this goal is the subject of a parallel RFA, HG-04-003.

The goal of research supported under this RFA is to develop or improve technology to enable rapid, efficient genomic DNA sequencing. The specific goal is to reduce sequencing costs by at least two orders of magnitude $100,000 serves as a useful target cost for a mammalian-sized genome because the availability of complete genomic sequences at that cost would revolutionise biological research and medicine. While not in a cost range that would enable the use of sequencing in individualised medicine, such technology would permit the sequencing of many genomes for a small fraction of current costs. A 100-fold cost reduction would make possible extensive studies of human variation for disease gene studies, substantially expanded comparative genomics to understand the human genome, and many other studies relevant to the National Institutes of Health (NIH), other federal agencies and the private sector. Entirely new lines of investigation would be enabled by making "large-scale sequencing" accessible to the diverse interests of many research laboratories and companies.

Many projects aimed at next-generation DNA sequencing technologies require substantial advances in a combination of fields such as signal detection, enzymology, chemistry, engineering, bioinformatics, etc. It is therefore anticipated that research programs responding to this RFA will involve multidisciplinary teams of investigators. The guidance for budget requests accommodates the formation of groups having investigators at several institutions, in cases where that is needed to assemble a team of the appropriate balance, breadth and experience. The scientific and technical challenges inherent in achieving a 100-fold reduction in sequencing costs are considerable. Achieving this goal may require research projects that entail substantial risk. That risk should be balanced by an outstanding scientific and management plan designed to achieve the very high payoff goals of this solicitation.

Although the ultimate goal of this RFA is to develop full-scale sequencing systems, independent research on essential components will also be considered to be responsive. However, it will be important for applicants proposing research on system components or concepts to describe how the knowledge gained as a result of their project would be incorporated into a full system that they might subsequently propose to develop, or that is being developed by other groups. Such independent proposals are an important path for pursuing novel, high risk/high pay-off ideas. Research conducted under this RFA may include development of the computational tools associated with the technology, e.g., to extract sequence information, including signal processing, and to evaluate sequence quality and assign confidence scores.

It may also address strategies to assemble the sequence from the information being obtained from the technology or by merging the sequence data with information from parallel technology. However, this RFA will not support development of sequence assembly software independent of technology development to obtain the sequence. The quality of sequence to be generated by the technology is of paramount importance for this solicitation. Two major factors contributing to genomic sequence quality are per-base accuracy and contiguity of the assembly. Much of the utility of comparative sequence information will derive from characterisation of sequence variation between species, and between individuals of a species. Therefore, per-base accuracy must be high enough to distinguish polymorphism at the single-nucleotide level (substitutions, insertions, deletions). Experience and resulting policy have established a target accuracy of not more than one error per 10,000 bases. All applications in response to this RFA, whether to develop resequencing or de novo sequencing technologies, must propose achieving per-base quality at least to this standard.

Assembly information is needed for determining sequence of new genomes, and ultimately also for genomes for which a reference sequence exists, to detect rearrangements, insertions and deletions. Rearrangements are known to cause diseases; knowledge of rearrangement can reveal new biological mechanisms. The phase of single nucleotide polymorphisms to define haplotypes is important in understanding and diagnosing disease. Achieving a high level of sequence contiguity will be essential to achieve the full benefit from the use of sequencing for individualised medicine, e.g., to evaluate genomic contributions to risk for specific diseases and syndromes, and drug responsiveness. Nevertheless, it is recognised that perfect sequence assembly from end to end of each chromosome is unlikely to be achievable with most technologies in a fully automated fashion and without adding considerable cost.

Therefore, for the purpose of this solicitation, grant applications proposing technology development for de novo sequencing shall describe how they will achieve, for about $1000, a draft-quality assembly that is at least comparable to that represented by the mouse draft sequence produced by December 2002: 7.7-fold coverage, 6.5-fold coverage in Q20 bases, assembled into 225,000 sequence contigs connected by at least two read-pair links into supercontigs [total of 7,418 supercontigs at least 2 kb long], with N50 length for contigs equal to 24.8 lib and for supercontigs equal to 16.9 Mb. The grant applications will he evaluated, and funding decisions made, in such a way as to develop a balanced portfolio that has strong potential to develop both robust re-sequencing and de nova sequencing technologies. If the estimate that achieving the goal of 100-fold reduction in cost for genome sequencing incorporating substantial assembly information will require about 5 years to achieve is correct, then re-sequencing technologies might be expected to be demonstrated in a shorter time. Grant applications that present a plan to achieve high quality re-sequencing while on the path to high quality de novo sequencing will receive high priority. The major focus of this RFA is on the development of new technologies for detection of nucleotide sequence. However, any new technology will eventually have to be effectively incorporated into the entire sequencing workflow, starting with a biological sample and ending with sequence data of the desired quality, and this issue should be addressed. Given that sample preparation requirements are a function of the detection method and the sample detection method affects the way in which output data are handled, these aspects of the problem are clearly relevant and should be addressed in an appropriate timeframe. However, NHGRI is interested in seeing that the most critical and highest-risk aspects of the project, on which the rest of the project is dependent, are addressed and proven as early as possible.

NHGRI anticipates that successful projects funded through this RFA may be sufficiently advanced as to be approaching early stages of commercialisation within about five years. Therefore, practical implementation issues related to workflow and process control for efficient, high quality, high-throughput DNA sequencing should be considered early. Some technology development groups lack practical experience in high throughput sequencing, and in testing of methods and instruments for robust, routine operation. Applicants may therefore wish to include such expertise as they develop their suite of collaborations and capabilities. The goal of this research is to develop technology to produce sequence from entire genomes. It is conceivable that sequence from selected important regions (e.g., all of the gene regions) could be determined in the near future, using more conventional technologies, at very low cost. However, that is not the purpose of this initiative, and grant applications that propose to meet the cost targets by sequencing only selected regions of a genome will be considered unresponsive.

This RFA will use NIH R21, R21/R33, R01 and P01 award mechanism(s). As an applicant you will be solely responsible for planning, directing, and executing the proposed project. Applicants may request an R01 or P01 (depending on the organisation of the proposed project) if sufficient preliminary data are available to support such an application. A fully integrated management and research plan should use the R01 mechanism. The P01 mechanism should be used if multiple projects under different leadership must proceed in parallel; however, the issue of synergy in a multi-focal effort is of great importance and must be addressed in the application.

Applicants requiring support to demonstrate feasibility may apply for either an R21 pilot/exploratory project or an R21/R33 award, which offers single submission and evaluation of both a feasibility/pilot phase (R21) and an expanded development phase (R33) in one application, The R21/R33 should be used when both quantitative milestones for the feasibility demonstration, and a research plan for the follow-on research, can be presented. The transition from the R21 award to the R33 award will be expedited by administrative review. The R21 alone is appropriate when the possible outcomes of the proposed feasibility study are unclear and it is not possible to propose sufficiently dear-cut and quantitative milestones for administrative evaluation, nor would it be possible to describe the R33 phase of the research in sufficient detail to allow adequate initial review.

This RFA uses just-in-time concepts. It also uses the modular budgeting as well as the non-modular budgeting formats. Specifically, if you are submitting an application with direct costs in each year of $250,000 or less, use the modular budget format. Otherwise follow the instructions for non-modular budget research grant applications. Applications must be prepared using the PHS 398 research grant application instructions and forms (rev. 5/2001). Applications must have a DUN and Bradstreet (D&B) Data Universal Numbering System (DUNS) number as the Universal Identifier when applying for Federal grants or cooperative agreements.

The Centre for Scientific Review (CSR) will not accept any application in response to this RFA that is essentially the same as one currently pending initial review, unless the applicant withdraws the pending application. However, when a previously unfunded application, originally submitted as an investigator-initiated application, is to be submitted in response to an RFA, it is to be prepared as a NEW application. That is, the application for the RFA must not include an Introduction describing the changes and improvements made, and the text must not be marked to indicate the changes from the previous unfunded version of the application.

Understanding a New Ingredient

In October 2010 the National Organic Standards Board recommended that engineered nanomaterials (ENMs) be prohibited from food products bearing the U.S. Department of Agriculture's coveted Organic label. If the department adopts the recommendation, ENMs will find themselves in the same officially taboo category as genetically modified organisms when it comes to organic foods nanotechnology-enabled innovations like flavour- and texture-enhancing ingredients and shelf life-extending packaging will be off the menu.

Prior to issuing its recommendation, the board received thousands of public comments and petition signatures supporting the ban and virtually none opposing it. Although an official decision could take years, supporters are confident the recommendation will be adopted, and it will go down as one of the first lines drawn in the sand when it comes to the reach of this relatively new and potentially transformative technology in the American marketplace.

Nanotechnology-enabled products are quietly proliferating on U.S. store shelves, despite nagging questions about the safety of synthetic nanoparticles and the products that contain them. "In our regulation of food and most consumer products, we don't implement the precautionary principle. Things go to market before we know whether or not they're really safe for human beings over the long term," says Alexis Baden-Mayer, a lawyer with the Organic Consumers Association, an advocacy group, who attended the meeting and campaigned for the ban.

Baden-Mayer and other observers perceive a distinct lack of public awareness about how common ENMs are becoming in the marketplace, and she hopes discussion among consumers of organic products will help change that. "Consumers don't know much about nanotechnology, and the first time they may hear about it is now when they learn that the organic regulations are going to prohibit [it]," she says.

The International Organisation for Standardisation defines a nanomaterial as a material with any external dimension between 1 and 100 nm. (By comparison, a double strand of DNA is about 2 nm thick.) Nanoparticles, which have been the focus of most nanotoxicology studies to date, one subset of nano-materials. Nanoparticles include structures of various shapes, such as nanotubes, nanowires, quantum dots, and fullerenes. They also occur naturally in substances like air, smoke, and sea spray, and "incidental" nanoparticles are created during processes such as combustion and food milling, churning, freezing, and homogenisation. (Naturally occurring and incidental nanoparticles were not included in the National Organic Standards Board's recommendation to ban ENMs.)

Nanotechnology the deliberate synthesis and manipulation of nanomaterials began in the 1980s. Today thousands of ENMs are manufactured in a kaleidoscope of substances, shapes, and sizes for use in a wide range of products and industrial processes that take advantage of their novel physical, thermal, optical, and biological properties. These properties may be determined by the ENM's chemical composition, size or shape, crystal structure, solubility, adhesion (the force that holds the nanoparticle components together), or surface chemistry, charge, or area.

Industry analysts have been forecasting "game-changing" advances as a result of nanotechnology in renewable energy, computers, communications, pollution cleanup, agriculture, medicine, and more. Clothing, sunscreens, cosmetics, sporting equipment, batteries, food packaging, dietary supplements, and electronics are just a few of the types of nanotechnology-enabled goods in use by U.S. consumers.

But safety questions arise around the nanoparticles in some of these products. The novel biological and physical properties of some ENMs pose unique challenges to comprehensive safety research, and investigators are working to figure out just how hazardous they might be to people, wildlife, and the environment. Compared with larger particles, nanoparticles' tiny size means tissues may take them up more readily. It also can give them an unusual ability to travel throughout the body, including into cells and cell nuclei, and across the placenta and the blood-brain barrier, as demonstrated in rodent studies. No cases of human illness or death have been definitively attributed to ENMs. However, a number of researchers and consumer and environmental advocates have warned that the abundant unknowns make it necessary to proceed with caution lest we repeat the history of asbestos, polychlorinated biphenyls, the insecticide DDT, and other innovations that seemed valuable when they were introduced, proceeded with little oversight, and ultimately caused major health or environmental problems.

What's on Store Shelves? As of 2007, the National Science Foundation estimated that up to $70 billion worth of nanotechnology-enabled products were sold in the United States annually, and that number is predicted to grow explosively. But pinpointing exactly which products contain ENMs is not always easy.

Manufacturers are not presently required to report the use of ENMs except for single-and multi-walled carbon nanotubes, for which the U.S. Environmental Protection Agency (EPA) finalised "significant new use" rules in September 2010. Another potential exception is the use of novel ingredients produced using nanotechnology in food or food packaging, which Sebastian Cianci, a U.S. Food and Drug Administration (FDA) spokesman, says "would in all likelihood require premarket approval." However, a report last year by the U.S. Government Accountability Office concluded that "FDA's approach to regulating nanotechnology allows engineered nanomaterials to enter the food supply as GRAS [generally recognised as safe] substances without FDA's knowledge."

Manufacturers also are not required to label products containing ENMs, and there seems to be a recent trend toward dropping voluntary references to such ingredients from packaging, and other publications, says Andrew Maynard, director of the Risk Science Centre at the University of Michigan School of Public Health. In some cases, he notes, "manufacturers will either just use the name of the chemical without stating whether it's at the nanoscale or not, or they'll use words like 'micronised,' so it's hard to work out whether it's nanoscale." The upshot, Maynard says, is that consumers are largely in the dark about whether the products they use contain ENMs.

In 2005 the Project on Emerging Nano-technologies (PEN) at the Woodrow Wilson International Centre for Scholars established an inventory of consumer products sold around the world that advertise having ENM content. The inventory now contains more than 1,000 entries. David Rejeski, PEN's director, says the inventory undoubtedly includes just a portion of the purportedly nanotechnology-enabled products on the market. Nonetheless, he says, the inventory has filled an important gap as the only catalogue of its kind, and it has enabled PEN to pick up several important trends that might otherwise have gone undetected. One was the rise of nano-silver, often used as an antimicrobial agent in products such as odour-resistant garments and food storage containers designed to keep leftovers fresh longer. PEN also spotted distinct upticks in products designed for children and babies and in products manufactured in China and other Asian nations.

However, funding for the inventory which came from the nonprofit Pew Charitable Trusts has run out, and PEN has neither added to the database since August 2009 nor been able to convince another institution to adopt it. Rejeski says that without an up-to-date inventory, researchers and regulators have no way to track the products available to consumers or the kinds of ENMs they may be exposed to. "It's not just a matter of the consumer not knowing," he says. "It's a matter of the government not knowing."

Products Under Scrutiny

That kind of information is important because when it comes to ENMs, different products lead to different potential exposures and therefore pose different potential hazards. For consumers, experts say, the greatest exposure probably comes from products that are ingested or otherwise come into intimate contact with the body things like dietary supplements, food, and personal care products.

The latter category includes such products as a hair growth-stimulating shampoo claiming to use copper nanoparticles, toothpastes with antibacterial silver nanoparticles, high-end skin cream made with "energising, detoxifying" gold nanoparticles, and "extreme wear" makeup. Sunscreens, however, have attracted the most attention from researchers and advocacy groups.

Many sunscreens include titanium dioxide or zinc oxide nanoparticles because they effectively block ultraviolet light while unlike the thick white creams of yore allowing the sunscreen to be transparent when it's rubbed onto the skin, according to the Environmental Working Group (EWG), a public health and environmental advocacy organisation. Testing of these products has focused on whether the nanomaterials can penetrate the skin, says Nigel Walker, deputy program director for science for the National Toxicology Program. As the saying goes, if there's no exposure, there's no risk, and several studies indicate that very little of the nanoparticles in sunscreen can penetrate the skin and enter the body as long as the skin is healthy and intact.

When it comes to sunscreens, Walker says, the research community is "pretty much comfortable that the amount of exposure for normal skin to nanoscale materials is extremely low, lower than many chemicals that we currently already use." Even the EWG which has petitioned the government to tighten regulation of nanomaterials in personal care products concluded that the risk of ultraviolet radiation damage from not wearing sunscreen outweighs the risk of harm from nanoparticles.

Still, concerns persist, especially since the lack of a rabelling requirement means people with skin abrasions or rashes, which are possible exposure pathways, could have trouble avoiding ENM-containing sunscreens. And people, especially children, are prone to consuming small quantities of sunscreen accidentally when they rub it onto their faces and lips. A lot of sunscreen also washes off in natural waterbodies or runs down the drain when people shower. Although more research is needed, initial studies have shown that titanium dioxide and zinc oxide nanoparticles can harm algae, , water fleas, , and frogs, and that they can travel up the aquatic food chain with unknown environmental consequences.

Titanium dioxide and zinc oxide ENMs, which are used in a huge array of products besides sunscreens, have been linked with potentially adverse health effects in some studies. For instance, a 2009 study from the University of California, Los Angeles, found that mice fed certain kinds of titanium dioxide nanoparticles with their drinking water for 5 days exhibited DNA and chromosomal damage and inflammation. In two separate studies the same year, a Japanese team showed that male offspring of pregnant mice injected with certain titanium dioxide nanoparticles experienced genital malformations and neurologic damage as well as changes in gene expression in the brain. Other in vitro studies have indicated some types of both titanium dioxide and zinc oxide nanoparticles are toxic to human brain and lung cells.

ENMs in edible products have garnered less research attention, the recommendation to keep them out of organic foods notwithstanding. One reason may be that it is unclear to what degree foods containing ENMs have actually hit U.S. supermarkets, although many observers have noted that the field of food science is unmistakably abuzz pursuing nanoscale ingredients to improve texture, taste, nutrition, shelf life, and safety. Cianci acknowledges that some salad dressings and spreads now on store shelves may contain nanoscale oil droplets intended to slow the separation of ingredients and that some fruits and vegetables may carry an edible coating of nanoscale wax droplets. "These are arguably foods that have engineered nanomaterials in or on them but which raise no safety concerns compared to their traditional counterparts," Cianci says.

As far as food packaging is concerned, the FDA is "not aware of any significant use of novel nanomaterials in the food packaging market at this time," according to Cianci. However, reports by groups such as PEN, the conservation organisation Friends of the Earth, and the British House of Lords indicate there is a great deal of research into using ENMs to develop advanced food packaging, and they point to several products already on the U.S. market, such as composite plastic bottles that incorporate nanoscale clays to extend the shelf life of beverages.

Dietary supplements are another market where nano is hot. The Source Vitamin Company, Inc., of Fort Lauderdale, Florida, for one, boasts of the All Natural Patented Nanotechnology driving its supplement products, which it maintains enables the targeted delivery and immediate absorption of active ingredients. How much truth there is behind such claims across the board is hard to assess because the FDA generally does not approve, test, or verify the rabelling of supplements before they hit the market. Andrew Shao, senior vice president of scientific and regulatory affairs at the Council for Responsible Nutrition, a trade group, says many supplement manufacturers' nano claims are simply "a marketing tactic." And Daniel Fabricant, vice president of global government and scientific affairs at the Natural Products Association, another trade group, says he doesn't know of any supplement manufacturers that are using nanoscale dietary ingredients they are simply too expensive, he says, and their benefits are unproven.

At the same time, Shao says some supplements do contain nanoscale ingredients that facilitate manufacturing processes or enhance properties such as the clarity of liquids. He points out that no adverse effects related to nanomaterials in supplements have been reported to the FDA, adding "there is no evidence that there's some kind of imminent threat as a result of the limited use and application of nanotechnology for supplements."

Still, the inclusion troubles many researchers, in part because of the lack of any premarket approval by the FDA. More important, the health effects of ingesting EN Ms remain poorly studied, researchers agree. "If you look at the data, virtually all of the tests [have] been done on exposure either through inhalation, through the skin, or through injection. There are virtually no studies on ingestion," says Michael Hansen, a senior scientist with Consumers Union, which has pressed for tight regulation of nanoparticles in consumer products. Research into the effects of long-term exposure to ENMs also is sorely lacking, he and other experts say.

Challenges to Health and Safety Research

Data on nanoparticle health and safety are hard won. Experts agree that many ENMs pose serious and unique scientific and methodological challenges to investigators, one very basic barrier being the ability to detect nanoparticles deposited in cells and tissues. For instance, it is exceedingly difficult to image materials smaller than 50 nm inside the human body, and quantifying carbon nanotubes is all but impossible, which poses a major challenge to assessing whether nanoparticles reach specific organs when evaluating data from toxicity studies. What instruments do exist can be prohibitively expensive, says Jaclyn Canas, a toxicologist with the Institute of Environmental and Human Health at Texas Tech University who has been studying the environmental fate of carbon nanotubes.

And that's only part of the problem. "It's a huge challenge to do nanotox-related research not just from a financial standpoint [but] from an intellectual standpoint, trying to wrap your mind around it and then really come to the bottom line of what causes toxicity," Canas says. "It's not as clear-cut as what we're trained to do and the kinds of contaminants we've worked with before."

Sheer numbers pose another basic but formidable challenge. Different coatings, sizes, surface charges, functionalisations, or manufacturing processes can drastically alter the toxicity and bahaviour of a given ENM, and companies are constantly developing new materials. By one estimate there are 50,000 different permutations of carbon nanotubes alone.

Moreover, there can be a great deal of inconsistency between batches. In one 2007 study Maynard led a team in analysing two batches of carbon nanotubes acquired from the same company. As is typical, the batches weren't pure: they contained distinct mixtures of single and bundled carbon nanotubes, unstructured carbon, and other metal nanoparticles. When agitated, however, one batch released dense particles of about 100 nm that were composed mostly of unstructured carbon, whereas the other tended to release larger particles that were diffuse, spiderweb-like tangles of bundled carbon nanotubes.

"You can imagine that the two different types of particles would behave completely differently inside the lungs," Maynard says. "So, notionally [they were] the same material, [but] in reality the types of particles being released were like night and day."

So great is the diversity and variation of ENMs that one report estimated conducting traditional in vivo toxicologic studies on the nanomaterials currently in commerce could take more than 50 years and cost upwards of $1 billion not to mention the sheer number of test animals required for such an endeavour. So there are several robust efforts under way to develop alternative testing protocols.

In one such effort, researchers in the EPA's ToxCast[TM] program are testing a small number of ENMs to see whether they are appropriate candidates for the program's high-throughput in vitro assays. Preliminary results indicate they are, says ToxCast leader Keith Houck, and this fall the EPA hopes to start bulk tests in earnest, focusing on various forms of ENMs including silver and titanium dioxide nanoparticles and carbon nanotubes. The goal is to use ToxCast data to priorities ENMs identified as hazardous for more detailed study and in vivo testing elsewhere.

Ultimately, asking the question "how dangerous are nanomaterials?" is likely to be fruitless, Maynard says. "There's no answer to that because some nanoparticles are going to be safe, some of them are going to be dangerous, some of them are just going to be very different. But if you ask a very specific question, like how dangerous are titanium nanoparticles of a certain size or a certain shape and what are we going to do about it, then you've got something that you can begin to apply scientific principles to."

Making Progress

Although progress began slowly, by most accounts significant strides have been made recently in developing detection devices, advancing research methodology, and accumulating both in vivo and in vitro toxicity data and the pace should only quicken. "There's so much more information coming out, particularly in the last year to two years," Walker says. "But the flip side is, it's a huge field, so it's always only a drop in the bucket."

Progress can't come fast enough for critics who accuse the federal government the chief funder of ENM health and safety research in the United States of selling this study area short. Fifteen different federal agencies conduct nanotechnology research, and their funding is reported through the National Nanotechnology Initiative (NNI). The NNI's 2010 research budget totalled an estimated $1.78 billion. About 5% of that was devoted to environmental, health, and safety research, with the rest going toward things like basic research into nanomaterial bahaviour, research facilities, and developing nanoscale devices and systems.

While pointing out that money for safety studies has increased every year and that study findings are starting to gel, Sally Tinkle, deputy director of the NNI's National Nanotechnology Coordination Office, concedes, "If there were more dollars we would move faster."

But Tinkle also says that, from the information accumulated so far, no acute health problems attributable to nanomaterials have been reported among workers, researchers, or consumers, although the effects of chronic exposure still require study. "It doesn't mean we should be less vigilant or we should slow down the pace of research, but it is a little bit reassuring," she says.

Despite all the uncertainties, many researchers express a similar sort of cautious optimism that nanomaterials will not follow the path of asbestos, polychlorinated biphenyls, and other harmful industrial agents. "I'm fairly confident that we're not going to have an epidemic of nanodiseases in twenty to forty years, similar to the epidemic of asbestos-related diseases that we had in the twentieth century," says Agnes Kane, a Brown University researcher whose rodent studies have shown that carbon nanotubes can produce effects very similar to those of asbestos fibres following instillation in the trachea or injection into the abdominal cavity. Maynard voices similar thoughts. "I think there is a greater chance that we're going to see long-term environmental impacts from these materials than we are going to see short-term consumer impacts," he says. At the same time, he cautions, "That is informed speculation because there are so many gaps in our knowledge."

Many researchers point to the emerging field known as green nanotechnology that is attempting to make ENMs and their production processes safer for people and the environment. They also look ahead to applications like the targeted delivery of chemotherapy drugs, tiny foodborne contaminant sensors, and advanced air- and water-filtration systems as plausible advances that could truly benefit society. But while many critics say they are enthusiastic about some of those positive applications, they remain adamant that safety research and regulation must catch up and keep up with the technology's proliferation. "I think we need to take a precautionary approach because we've learned the hard way over and over and over again," says Hansen. "You'd think we would learn."

Chapter 6

Analysis of Bronchoalveolar Lavage Fluid

In recent years low-molecular-weight serum protein profiling has become increasingly important in detecting early events in the disease process and predicting outcomes. The recent use of this technique to detect ovarian cancer provided a great impetus to this field. Both SELDI-TOF (surface-enhanced laser desorption/ionisation-time of flight) and MALDI-TOF (matrix-assisted laser desorption/ionisation-time of flight) mass spectrometry are proving to be powerful tools for diagnosing disease states, particularly for early detection of cancer, through the analysis of proteomic patterns. When combined with bioinformatic tools, protein profiling has become an effective approach in screening for potential tumour markers. Because bronchoalveolar lavage fluid (BALF) exhibits the cellular and biochemical alterations of inflammation and lung injury in response to various toxic agents, performance of a proteomic analysis of BALF to characterise the effects of diesel exhaust particles (DEPs) exposure is warranted. We previously designed a neural network program for the analysis of proteomic patterns in serum samples of humans exposed to various levels of DEPs (unpublished data). These studies showed the potential for proteomics to discriminate occupational exposures to various deleterious agents and prompted the validation study presented here.

DEP exposure induces the production of cytokines in lung epithelial cells in vitro and in lung tissue in vivo. It also affects the lipopolysaccharide-induced production of cytokines (tumour necrosis factor-[α] and interleukin-1) in alveolar macrophages. We previously studied the expression of the mRNA levels for several of these cytokines and correlated these observations with the inflammatory response as assessed by measuring the influx of cells and protein into the bronchoalveolar space. In addition, cytokine levels were measured in BALF. The results showed that DEPs up-regulate several genes implicated in the inflammatory response, both at the message and protein levels, within 24 hr in cells obtained from BALF, representing the influx of both polymorphonuclear leukocytes (PMNs) and AMs.

In this study, we used newly available proteomic technologies to characterise the changes in protein concentrations caused by DEP exposure. We used a Ciphergen Protein Chip System and liquid chromatography coupled to mass spectrometry (LC/MS) to characterise the samples.

In the Ciphergen system, protein samples are allowed to adsorb to spots on a fixed support with a specific surface chemistry. Unbound proteins are washed off the chip, and the remaining bound proteins are ionised with a laser, and their masses are characterised by time of flight mass spectrometry. For LC/MS, polypeptide mixtures are digested with trypsin; the peptides are bound to a chromatographic column, eluted with a continuous gradient of acetonitrile and ionised by electrospray directly into either a time of flight or ion-trap mass spectrometre. Using a weak cationic exchange Protein Chip, protein profiling was performed on BALF taken from rats at 1, 7, or 30 days after exposure to various concentrations of DEPs. This approach was complemented by global analysis using LC/MS to determine protein identity and to broadly screen for qualitative differences. We found DEP exposure-induced changes in the abundance of a number of proteins using a SELDI methodology. These and additional proteins identified by LC/MS are indicative of tissue damage and inflammation.

Research was conducted in compliance with the Animal Welfare Act (1966), and other federal statutes and regulations relating to animals and experiments involving animals and adheres to principles stated in the Guide for the Care and Use of Laboratory Animals (National Research Council 1996) in facilities fully accredited by the Association for the Assessment and Accreditation of Laboratory Animal Care, International. The animals were treated humanely and with regard for alleviation of suffering. The animals used in these experiments were specific pathogen-free male Sprague-Dawley rats, weighing 250-275 g (approximately 8 weeks old) at arrival. The rats were housed at the National Institute for Occupational Safety and Health animal facility, under temperature and humidity controlled conditions and a 12 hr light/dark cycle. The rats were monitored to be free of endogenous viral pathogens, parasites, mycoplasmas, Helicobacter, and CAR (cilia-associated respiratory) bacillus. Rats were acclimated for at least 5 days before use and were housed in ventilated cages, which were provided with HEPA-filtered air. Alpha-

Dri virgin cellulose chips and hardwood Beta chips were used as bedding. ProLab 3500 diet and tap water were provided ad libitum.

Reagents

DEPs were from a National Institute of Standards and Technology standardised heavy-duty diesel engine emission sample (no. 1650) with an average mass median diametre of 0.5 [micro]m.

Experimental Design

Animals were exposed by intratracheal (IT) instillation with a single dose of either saline, or DEPs. Groups of animals (n = 4 per group), representing each treatment, were sacrificed at 1, 7, and 30 days after exposure to obtain BALF.

Intratracheal Instillation of DEPs

DEPs were suspended in endotoxin, [Ca^{2+}] and [Mg^{2+}] free phosphate-buffered saline and sonicated for 1 min. Rats were anesthetized with an intraperitoneal (ip) injection of 30-40 mg/kg body weight sodium methohexital and were intratracheally instilled using a 20-ga, 4-inch ball-tipped animal feeding needle. Rats were given 5, 35, or 50 mg DEP/kg body weight or an equivalent volume of PBS. IT instillation has been shown to be a valid model to study pathological changes associated with airborne pollutants and is considered particularly useful in elucidating mechanisms of response.

BAL Fluid

Rats were anesthetized with an overdose of sodium pentobarbital (100 mg/kg body weight) and exsanguinated. The trachea was cannulated, and the lungs were lavaged. BALF was obtained by a single lavage using cold [Ca^{2+}]- and [Mg^{2+}]-free PBS containing 5.5 mM D-glucose. The first lavage return of approximately 6 mL was centrifuged to sediment cells at 300 x g. The acellular supernatant was transferred to plastic tubes and stored at -80°C. Every effort was made to minimise protein degradation by avoiding multiple freeze-thaw cycles. Samples analysed by SELDI-TOF were thawed only once. At times additional processing of samples, such as weak cation exchange (WCX)-extraction, required refreezing of aliquots.

Proteomic Patterns

Proteomic patterns were obtained using the WCX2 ProteinChip on the Ciphergen ProteinChip System (Ciphergen Biosystems, Inc., Fremont, CA). WCX2 chips were equilibrated with 2x binding buffer

[50 mM ammonium acetate (N[H_4]OAc) and 0.01% Triton X-100 at pH 6.0]. BALF samples were diluted 2-fold with binding buffer, and 200 [micro]L was placed on a WCX2 chip in the bioprocessor and allowed to incubate for 1 hr. Chips were washed with binding buffer and water before drying and the addition of the sinapinic acid (SPA). Data collection was optimised for the mass to charge ratio (m/z) range of 3,000-50,000, with a detector sensitivity of 7, a laser intensity of 150 and a high m/z of 50,000. The data presented in the figures have been baseline subtracted using Ciphergen's ProteinChip software with a window of 25 points for the option to smooth before fitting baseline and with the automatic option for parameters.

Protein Identification

Protein samples from a control rat and a rat 24 hr postexposure to the high dose were subjected to further analysis to identify the proteins associated with the peaks observed in the SELDI-TOF data. To mimic WCX2 chemistry, a solid-phase extraction (SPE) was performed with the BALF using a WCX resin, Biosepra CM Ceramic HyperD F. The resin was equilibrated with 1x binding buffer. BALF was diluted with 2x binding buffer (1x final concentration) and mixed with the resin. The resin with bound BALF proteins was washed with 0.5 M N[H_4]OAc to remove proteins with a low binding affinity, then proteins of interest were eluted with 2 M N[H_4]OAc. Eluted proteins were separated using a NuPAGE 4-12% BisTris gel in an MES Running Buffer and stained with SyproRuby. Protein bands in the appropriate relative molecular weight range were manually excised, reduced with 5 mM dithiothreitol (DTT) and alkylated with 50 mM iodoacetamide (Bio-Rad Laboratories), digested with trypsin (Promega Corp., Madison, WI) and eluted by diffusion. After concentration by evapouration in a Speed Vac Concentrator, samples were resuspended, and an aliquot of each elution was modified with imidazole to enhance ionisation. Imidazole was part of the Lys Tag kit and used according to manufacturer's recommendations. Both rabelled and unrabelled samples were analysed on an Agilent SL ion trap mass spectrometre connected to an Agilent 1100 nanoflow HPLC (Agilent Technologies).

As a secondary validation of identification, WCX-extracted proteins and whole BALF were analysed by direct LC/MS (not gel based) on a Waters Q-Tof Premier quadrupole time-of-flight mass spectrometre (QTOF) in tandem with a nanoACQUITY ultra performance liquid chromatograph (UPLC) system. Before LC/MS, the samples were lyophilised and resuspended in 50 mM ammonium bicarbonate, 5 mM

DTT and 0.05% RapiGest. After denaturation at 80°C, proteins were alkylated with 25 mM iodoacetamide and digested overnight with trypsin. The samples were diluted to a final RapiGest concentration of 0.025%. For the whole BALF, a tryptic digestion of the proteins was separated into eight fractions using an Agilent 1100 HPLC before analysis on the QTOF. In lieu of traditional tandem MS (MS/MS), QTOF data were collected using the Waters Protein Expression method. Three technical replicates were performed for each of the samples.

Protein Expression Method

This is a novel method of data acquisition and analysis designed by Waters to maximise information content gained from mass spectral analysis. In this method, spectra are collected alternating between low- and high-collision energies; no selective mass filtering is performed. Therefore, fragmentation data are collected for every precursor ion and are not limited by the number of MS/MS scans that can be performed in a single run. Furthermore, the intensity of each precursor ion is collected across its entire peak, so quantitative data are maximised. The fragment ions in the high-energy scans are assigned to precursor ions based on elution profiles using computational methods. The collection of fragment ions is combined into a synthetic spectrum (termed M[S^E], where E signifies energy) that is used for database searches.

LC Parameters

Peptides extracted from gel spots were separated using an Agilent 1100 HPLC coupled to the Agilent SL ion trap mass spectrometre. The 8-[micro]L injection volume was trapped on a 0.3 x 5-mm Zorbax SB C-18 column using 3% acetonitrile (MeCN) and 0.1% formic acid at a flow rate of 20 [micro]L/min for 16 min. A 75-[micro]m x 50-mm Zorbax SB C-18 column, 3.5-[micro]m particle size (Agilent Technologies), was used for analytical separation with a flow rate of 300 nL/min. The gradient profile was 3% MeCN for 16 min, 10% MeCN at 23 min, 35% MeCN at 43 min, 80% MeCN at 48.5 min until 58.5 min, and 3% MeCN at 63 min until stopping at 67 min; 0.1% formic acid was used throughout.

Initial fractionation of whole BALF peptides was performed using a combination of anion and cation exchange columns on an Agilent 1100 HPLC. A Polycat A 200 x 4.6 mm, 5 [micro]m, 300-[Angstrom] column and Polywax LP 100 x 4.6 mm, 5 [micro]m, 1,000-[Angstrom] column (PolyLC Inc., Columbia, MD) were connected in series for the separation. The injection volume was 100 [micro]L and the column temperature was 35°C. The gradient profile was from 20 mM N[H_4]OAc to 1.8 M N[H_4]OAc in 9 min and held constant until stopping at 17 min; 10% MeCN was used throughout. Time-based fractions were collected starting at 2.1 min. A 1-min fraction and four 30-sec and three 1-min fractions were collected in order. Samples were dried and resuspended in 100 [micro]L of 3% MeCN, 0.1% formic acid.

WCX-extracted BALF and fractionated whole BALF were separated using the Waters nanoACQUITY UPLC system coupled to a Q-Tof Premier mass spectrometre. The injection volume of 10 [micro]L was trapped using a 180 [micro]m x 20 mm Waters Symmetry C18, 5-[micro]m particle size column using 0.1% formic acid at a flow rate of 5 [micro]L/min for 4 min. The analytical separation was performed using a 75 [micro]m x 100 mm Waters nanoACQUITY UPLC BEH C18 column, 1.7-[micro]m particle size. The column temperature was 35°C. The gradient profile was 3% MeCN for 1 min, 30% MeCN at 101 min, 60% MeCN at 105 min, 80% MeCN at 111 min, and 3% MeCN at 112 min until stopping at 130 min; 0.1% formic acid was maintained throughout. The flow rate was 300 nL/min for the extracted BALF samples and the first fraction of whole BALF. The flow rate was reduced to 250 nL/min because of high back pressure for the remaining fractions.

Ion Trap Parameters

Peptides were ionised in positive ion mode. Ion charge control was used with a target of 75,000 counts and a maximum accumulation time of 300 msec. Three precursors were selected based on intensity with an absolute threshold of 1,000 counts. Active exclusion was used and precursors were released after 1 min. The MS/MS fragmentation amplitude was set at 1.2 V.

QTOF Parameters

Peptides were ionised in positive ion mode. Data were collected over the m/z 50-1,900 range for 0.8 sec/scan. Scans were performed with the collision cell voltage set at 10 V for low-energy scans and ramped from 20 to 40 V during high-energy scans. [[Glu^1]]-fibrinopeptide B was used as an external lock mass for accurate mass calculations (m/z 785.8426). A 1-sec lock mass scan was collected every 30 sec.

Ion trap data were converted to peak lists using Data Analysis 2.2 (Agilent Technologies). Mascot 2.1 (Matrix Science, Boston, MA) was used for searching with following modifications enabled: fixed carbamidomethyl (C) and variable oxidation (M), oxidation (HW), phospho (STY), sodiated (C-term), and sodiated (DE). A variable imidazole modification was also enabled when appropriate. Under these conditions, the 95% confidence level for an individual peptide match corresponded to Mowse scores ranging from 50 to 54.

The QTOF data were submitted as raw data to Protein Lynx Global Server 2.2 and processed using the Protein Expression method. The only modification enabled in PLGS searches was a fixed carbamidomethyl (C). To limit the number of false positives, we considered only protein identifications with confidence levels [greater than or equal to] 0.99. When multiple isoforms of the same protein were identified that shared numerous identical peptides and could not be distinguished, the confidence levels were summed, and a single protein was reported with multiple accession numbers. Protein identifications determined by direct LC/MS were reported only if they were found in all three technical replicates of at least one condition and had identifications for at least three unique peptides. Despite these rigid criteria, we found that when combining the peak lists from all the LC/MS runs from the fractionated BALF, there was an unacceptably high false positive rate (as determined by the number of random sequence hits). The scoring algorithm used in PLGS appears to overestimate the relevance of numerous, low-quality hits. The effect is pronounced only in large data sets, where presumably the total number of peaks increases the chances that multiple incorrectly identified peptides may be attributed to the same protein in the database. To control for this, we developed a second scoring criterion using the average score per peptide and the average score for the top five highest scoring peptides. Only protein identifications with an average score per peptide > 2.7 or average of the top five peptides > 10 were considered to be high-quality hits. These values were set at a level that limited the false positive rate to < 5% in single replicates and allowed no detectable false positives when the replicate filter was used.

When compared with the mass spectra from the control animals, the spectra from the DEP-exposed animals show two additional peaks with m/z values of approximately 9,100 and 10,100. These peaks appear in samples from all the exposed doses at 24 hr but are not seen in day 7 and day 30 samples. This finding indicates that these peaks represent an acute response to DEP exposure that resolves in a few days. To identify these proteins, we fractionated BALF from rats 24 hr postexposure by SPE using a WCX resin with subsequent denaturing polyacrylamide gel electrophoresis (SDS-PAGE). Four predominant bands are observed in exposed samples in the low-molecular-weight region of the gel where the proteins corresponding to the peaks of interest from the SELDI-TOF data would be expected to migrate. These gel bands were excised; the protein contained in them was digested, eluted, and analysed by LC/MS technology using an ion trap mass spectrometre. The uppermost band is present at approximately the same concentration in both exposed and unexposed samples and was identified as lysozyme. The two lowest bands likely correspond to the SELDI-TOF peaks with m/z 9,100 and 10,100 and were identified by database searches as anaphylatoxin C3a and calgranulin A. We could not identify the protein from the middle band by analysis of the excised gel slice. A small peak corresponding to its relative molecular weight is detectable in the SELDI-TOF data with an m/z just above 13,000. However, it is only distinctly above noise level when multiple spectra are averaged.

Two predominant peptides from the SDS-PAGE-excised bands formed the basis for the identification of anaphylatoxin C3a, which is a proteolytically processed product of complement C3. Two peptides present in complement C3 but not in anaphylatoxin C3 were identified, suggesting that complement C3 or partially cleaved C3 might be present. However, we believe that the identification of these peptides as C3 sequences is artifactual because they were the two lowest scoring peptide matches in the search (scores = 3 and 6), their scores were below statistical significance, and they were only found in imidazole-rabelled samples, even though neither peptide has this modification. We conclude that the presence of naturally processed anaphylatoxin C3a is the most likely explanation for the presence of complement C3 peptides.

SPE-fractionated BALF was also analysed directly by LC/MS (no SDS-PAGE) using a QTOF and the Waters Protein Expression method, and the presence of calgranulin-A was confirmed. Complement C3 or a C3 isoform, 98% identical to C3 overall and 100% identity in the anaphylatoxin C3a region) was also identified, but since proteins are digested with trypsin before LC/MS analysis, the full-length and processed proteins are indistinguishable. Additionally, this technique provided a possible protein identification for the previously unidentified band and the small SELDI-TOF peak at 13,000, calgranulin B.

A total of 65 proteins were identified by performing LC/MS analysis of whole BALF and WCX-extracted BALF on a QTOF using the Waters protein expression method. Each reported protein was identified in all three technical replicates of at least one of the conditions. We compared the lists of confirmed proteins and unfiltered search results to identify possible missed or lower scoring identifications. The quality of the identification of each protein in each condition was assigned in one of four ranks: a) high-quality identifications in all three replicates, b) a high-quality identification in at least one replicate, c) a low-quality identification in at least one replicate, and d) not identified. The majority of the proteins was seen in both control and exposed samples, whereas 20 were identified only in diesel-exposed samples and 4 were identified only in control samples.

The predominant peaks found in the SELDI-TOF spectra have all been identified. Lysozyme (Lyz) is quite abundant in these samples and appears as singly, doubly, and triply charged peaks (m/z 15,000, 7,500, and 5,000, respectively). Two of these are the largest peaks in all the samples and do not change as a result of exposure. Two readily observable peaks, seen only in spectra from the diesel-exposed, 24-hr samples, represent anaphylatoxin C3a and calgranulin A, and a third peak, whose signal is only slightly above noise, is also present only in spectra from exposed, 24-hr samples and was identified as calgranulin B.

Diesel exhaust particles are generated by heavy-duty diesel engines used in many industries and motor vehicles used in public transportation. They are respirable particles with an average diametre of 250 nm and contain several mutagenic and carcinogenic hydrocarbons. Epidemiologic and experimental animal studies have shown an increased risk of respiratory and cardiovascular morbidity and mortality associated with exposure to DEPs. They also cause adverse reactions in the lungs and other tissues. DEP exposure induces production of cytokines in AMs, in lung epithelial cells, and in lung tissue. Within 24 hr after exposure to DEPs, quantifiable changes in cytokines, an influx of inflammatory cells and proteins, and up-regulation of gene expression of inflammatory mediators are observable in BALF from exposed rats.

The aim of this study was to characterise the changes in protein profiles in the BALF of rats following DEP exposure using newly available proteomic technologies. This was accomplished using two complementary technologies, SELDI-TOF and LC/MS. The spectra obtained using a Ciphergen ProteinChip System contains two readily observable peaks and a third weak peak that are specific to BALF samples taken from rats 24 hr after exposure to DEPs. Subsequent analysis using LC/MS indicates that the proteins producing these peaks are calgranulin A, calgranulin B, and anaphylatoxin C3a.

An additional 62 proteins present in BALF were also identified through the utilisation of Waters's LC/MS protein expression technology. Twenty proteins, most of which are lung damage and inflammation specific, were repeatedly identified only in the exposed sample. Presumably, they are more abundant in this sample, as there is a strong bias toward identification of the proteins at the highest concentrations using LC/MS. Four proteins were identified in the control sample but not in the exposed one. The abundance of these proteins may have been reduced in BALF from exposed animals. However, the presence of these proteins may also simply have been masked in the exposed sample by the higher amount of total protein in it.

There is a large difference in total protein levels between the two samples, with the higher protein concentration in the exposed sample likely a result of plasma extravasation. Consistent with this view, many of the plasma-derived proteins identified in both samples do indeed change in abundance, but additional work will be required to provide accurate quantification. A quantitative comparison using the Protein Expression method was confounded by ion-suppression effects and challenges in normalisation resulting from the large difference in total protein concentrations. Because we prefer to interpret the data conservatively, we have not reported this "quantitative" data.

A side note to the LC/MS analysis is that different sets of proteins were identified in the SPE samples and in the whole BALF. The largest hindrances to protein identification by mass spectrometry are sample complexity and dynamic range. The WCX extraction addresses both of these by reducing the concentration of the most abundant protein (albumin) and reducing the total number of proteins present. This is why proteins that are "hidden" in whole BALF can be identified in SPE samples.

Based on the proteins identified, the major observed effect of DEP exposure appears to be an inflammatory response. Anaphylatoxin C3a, a component of the complement system, is a well-known mediator of inflammation, and calgranulin A is a part of a hetero-dimer with calgranulin B (also known as MRP-8 and MRP-14, respectively) that makes up calprotectin. Calprotectin is currently used as a biomarker of inflammation for several human diseases. It is primarily expressed in PMNs and is estimated to account for 30-60% of their cytosolic and 5% of their total proteins. Furthermore, the level of calprotectin at sites of inflammation is known to correlate with the number of localised PMNs [reviewed in Striz and Trebichavsky (2004); Yui et al. 2003], and we have previously shown that exposure to DEPs causes an increase in the PMN content of BALF at 24 hr. The presence of an inflammatory response is further supported by the qualitative analysis of the proteins identified by LC/MS.

Many of the proteins that we have observed in both the exposed and unexposed sample are highly abundant in the plasma and are present as a result of plasma extravasation. However, DEP-exposed samples show a pronounced increase in the amount and number of proteins observed, which appears to be caused by damage at the air-blood barrier that is a result of DEP exposure. Without analysis of the plasma, it is not possible to discriminate changes in concentration of plasma-derived proteins that are present because of extravasation from those that are specific to the inflammatory responses.

On the technical side, it is worth noting that the discovery of anaphylatoxin C3a in the SELDI-TOF data demonstrates an advantage to a top-down proteomics approach. LC/MS analysis of digested proteins is unable to distinguish processed and unprocessed complement C3. The SDS-PAGE analysis confirmed the SELDI-TOF data but required significantly more effort to acquire the data. Protein modifications, such as the cleavage of complement C3, can play major biological roles and are important to characterise. However, it must be noted that the top-down approach used in this work is not a global analysis, as only proteins that bind with high affinity to a weak cation exchange chip were retained. The WCX2 chip was chosen because of its low affinity for serum albumin, which is abundant in BALF; in the future the analysis could be extended by using chips supporting other surface chemistries. Another caveat was that this approach only detected large acute changes and was unable to discern difference at later time points. Since it is known that the lungs have not returned to normal, it is likely that the method was not sensitive enough or does not have sufficient dynamic range to identify the lower abundance proteins that are perturbed at 7 or 30 days postexposure. In the current state of proteomic technology, a combination of complementary strategies is required to maximise proteome coverage.

The protein expression method used in this study provides more complete coverage than the SELDI-TOF and was chosen over a traditional LC-MS/MS approach because it is more suited for comparative analyses. This method of fragmentation provides more reproducible coverage because all the ions are fragmented each run, and the identified peptides are not limited by which precursor ions are select for tandem MS. Furthermore, since only one fragmented ion scan is performed for each survey scan, there is far more quantitative data for the parent ions. This being said, two noteworthy shortcomings were identified with this approach.

The scoring algorithm used for database searches needs to be optimised. We addressed the issue by removing proteins that were identified as the result of multiple low scoring peptide hits, but ideally such filtering should be handled in the initial search. There is no easy way to normalise or to account for ion suppression resulting from samples with large difference in protein abundance, such as the unexposed and exposed BALF samples. In summary, we demonstrate that it is possible to detect markers of inflammation after diesel exhaust particulate exposure in the BALF using the Ciphergen ProteinChip System. Additional mass spectrometric investigation using liquid chromatography coupled to mass spectrometry (LC/MS) was used to identify the predominant peaks present in the SELDI-TOF spectra and also provided an additional list of proteins that change in response to exposure. Further studies are required to see if these markers are detectable in serum samples from animals or humans exposed to diesel exhaust.

Species Composition and Structure of Vascular Plants and Bryophytes

Calcareous tufa-forming species-rich low-productive fens are among the most endangered wetland types in the world. This fen type can be characterised by precipitation of the calcite as consolidated tufa or unconsolidated marl on the surface. The vegetation consists of many calcifuge specialist species that favour the low content of available phosphorus in a high-calcium substrate. Such an environment can develop in the conditions of a stable near-surface water level by recharge of calcium-rich groundwater. A drop of the water level and an increase in its seasonal fluctuation result in complete desiccation of the topsoil. This influences the redox status the nitrogen availability increases, but phosphorus availability does not increase because of limited decomposition. Electrical conductivity decreases as recharged groundwater dilutes with precipitation but pH remains stable until the bicarbonate buffer system functions. The water level drawdown is responsible for the decrease in plant species diversity as many specialist species are out competed by a few generalists like Molinia caerulea and for the encroachment of woody species.

Rich fens have been used for haymaking and cattle grazing in Western Europe for a long time. Among other reasons the decrease in species richness on these undrained or slightly ameliorated fens is argued to be related with the cessation of the traditional management regime, which is responsible for increasing litter accumulation.

Mineral-rich (mainly calcium, but also magnesium and iron) groundwater-fed fens are spread in regions where calcareous sediments lie close to the surface. In Estonia, calcareous-rich fens are concentrated in the western and northern areas and on the western island of Saaremaa, where the Silurian carbonate rock and Ordovician limestone and marls are covered by thin Quaternary sediments. Estonian fens have been mown and grazed by cattle, drained, and turned into pastures, hayfields, and crop fields since the middle of the 17th century. Since the 1950s the area of intact rich fens decreased from about 80 500 ha to 5000-8000 ha in the 1990s. The rich fen sites left in the near-natural state are mostly fragmentary and even protected fens are under indirect drainage impact, although in many cases the characteristic species composition is largely preserved.

This study was conducted at a calcareous fen site of the Paraspollu Nature Reserve (also a Natura 2000 site) in North Estonia, which was in part slightly drained and used for mowing and cattle grazing for a long time (at least from the end of the 19th century), but was abandoned in the early 1960s. The site is important because of its rich flora with many rare vascular and moss species that are protected regionally. In all, 19 rare and protected vascular plant species are known to inhabit the Paraspollu Nature Reserve.

Our objective was to identify factors associated with the composition and structure of plant communities within the Paraspollu calcareous-rich fen site. The more detailed purpose was to determine the relationships between vascular and moss species coverage, water table position and its seasonal fluctuations, water pH and electrical conductivity (EC), and litter, but also microtopography (coverage and height of tussocks) and coverage of Molinia caerulea. Three transects were established over the fen site, and vegetation, microtopography, water table position, and water pH and EC data were collected from 23 releves of 1 m x 1 m. This pilot study is the first step towards finding effective ways of management that could help to decrease the predominance of the few ubiquitous species that evidently are outcompeting specialist species characteristic of calcareous-rich fens.

The Paraspollu Nature Reserve (253 ha) constitutes the south-eastern part of the Peningi mire complex (7552 ha), which lies on the Harju plateau in North Estonia. It is surrounded by glaciofluvial ridges in the west and south and by a till plateau in the east and north. Fen communities are distributed in the southern and eastern parts of the mire complex. The gradient in moisture conditions and vegetation pattern can be well distinguished over the open calcareous fen site (ca 40 ha) (59°18' N, 25°7' E) selected for the study.

The peat deposit is formed of reed-sedge and reed-brown moss fen peats with a medium degree of decomposition, and is underlain by tufa in some parts of the site. The thickness of the peat layer reaches 3.2 m and that of tufa 1.1 m, underlain by a sandy-clayey till deposit.

Four hand-cut ditches bordering and crossing the fen site can be found on a 'One Verst Map' (one verst = 1066.788 m) from as early as about 1900. A denser drainage network is marked on the map from the mid 1930s. In the 1950s the site (open fen) was bordered by deeper ditches in the west and south-west, but these are partly filled with sediments and the actual depth of most ditches is 0.2-0.5 m. The fen surface gradient and water flow are directed roughly to the north-west. The density of the drainage network of the area, including border ditches, is 106 m [ha^{-1}].

The study site is open calcareous-rich fen with a sparse tree layer (mainly Betula pubescens and Picea abies), surrounded by paludified forest (in the east) and slightly drained bog pine forest (in the west). Trees and shrubs (Pinus sylvestris, Picea abies, Betula pubescens, B. humilis, B. nana) invade into the fen area mainly from its borders. The tree canopy is denser in the western and southwestern parts of the fen close to deeper border ditches.

Vegetation Sampling

In June 2007 the composition and coverage of vascular plant species of the field layer and bryophyte species of the surface layer were recorded in 23 releves 1 m x 1 m in size along three transects. As one plot contains on average 22% of the site's vascular plant flora and some 16% of the moss flora, the size of the plot is adequate for ordination analyses. Additionally, the total species list of the area was compiled. The nomenclature of bryophytes follows Ingerpuu & Vellak (1998) and that of vascular plants, Leht (2007). The coverage of litter and the height and coverage of tussocks were estimated. The coverage of vascular plants and bryophytes was recorded using the percentage scale.

Pore Water Sampling

Depth to water level (DWL) from the ground surface between tussocks was measured (30 October 2007, 30 May and 30 July 2008), and the pore water samples were taken (1 November 2007, 1 June 2008) from piezometres (PVC tubes with the inner diametre of 22 mm, perforated by 10 cm and inserted into the surface to a depth of 50 cm) close to each releve. Samples were taken to the laboratory, kept at room temperature for 24 h, and pH (Handylab pH11/SET, SCHOTT Instruments GmbH) and EC (Conductivity Metre Micrometre 900) were measured.

The Mantel test was used to evaluate the null hypothesis of no relationship between the field and surface layers. Standardised Mantel statistic (r) was calculated using Mantel's asymptotic approximation method. The size of field and surface layer dissimilarity matrixes was 43 species by 23 plots and 24 species by 23 plots, respectively.

The classification and ordination of field and surface layer plant cover data were performed by PC-ORD software. Based on two-way cluster analysis of species cover data for all plots, five plant assemblages were identified. We used the Sorensen (Bray-Curtis) distance measure and the Flexible Beta group linkage method with the flexible beta value of -0.250. Of vascular plant species 13 and of moss species 20 were exluded from the matrix as rare species (found per one plot). The final data matrix consisted of 23 plots and 43 vascular plant (herbs, shrubs, and trees in the field layer) and 24 moss species. Nonmetric Multidimensional Scaling (NMS) ordinations using random starting points were used to explore the relationships between the vegetation and environmental data. Data transformation by relativisation by column maximum was performed to improve the assumption of normality and homogeneity of variance and to make units of environmental variables measured in different scales more comparable. This adjustment also tends to equalise common and uncommon species.

The proportion of variance represented by each axis was found based on the [r^2] between the distance in the ordination space and the distance in the original space. In order to interpret the ordination, multiple correlations were calculated between ordination axes and the plant species coverage, and between the axes and each environmental variable.

Species Density and Coverage of Species Groups

Altogether, 44 species of bryophytes (incl. 9 hepatics) and 70 (57 in releves) vascular plant species were found on the study site. Lichens were not identified to species level. The total list contains 13 rare species, among others 11 in vegetation releves. Species richness per plot was 19.3, evenness 0.50, Shannon's diversity index 1.46, and Simpson's diversity index for infinite population 0.65. The species density of bryophytes varied between 4 and 12, of graminoids between 4 and 14, and of woody species (trees, shrubs, and dwarf shrubs) between 0 and 4 per plot.

Description of Clusters

Using two-way cluster analysis, we identified five groups from 23 plots. Cluster 1, the Phragmites australis Calliergonella cuspidata group, occurred only in two plots (2 and 3) in the northern part of the area. It consisted of 4 species in the shrub layer, 15 species in the surface layer, and 8 species in the ground layer. Surface layer coverage was not high; Carex lasiocarpa, Phragmites australis, and Menyanthes trifoliata dominated. In the ground layer Drepanocladus cossonii, Calliergonella cuspidata, and Campylium elodes covered most part of submerged depressions between tussocks. The depth of the water level was rather stable close to the surface and fluctuated up to 5 cm between seasons.

Cluster 2, the Carex panicea Schoenus ferrugineus Drepanocladus cossonii group, was found in five plots (1, 4, 11, 12, and 19) and was distributed around cluster 1. This was the most species-rich cluster distinguished. The releves of the cluster were represented by 5 species in the shrub and tree layers, 27 species in the surface layer, and 18 species in the ground layer. Schoenus ferrugineus dominated in the ground layer with Carex davalliana, C. panicea was the codominant. Selaginella selaginoides was found on a tussock in two releves. The depressions between S. ferrugineus, C. davalliana, and M. caerulea tussocks were well covered by mosses, with Drepanocladus cossonii dominating with mean coverage of about 19% and Campylium stellatum, Meesia triquetra, and Scorpidium scorpioides being other more common species. At the end of October the depressions between tussocks were submerged. In May and July the water level was close to the surface (plots 11, 12, 19) or at a depth less than 20 cm (plots 1 and 4).

Cluster 3, the Molinia caerulea Carex davalliana group, was found in four releves (13, 16, 18, and 20). Tussocks covered some 20% and litter some 70% of the surface. We found 4 species in the shrub layer, 19 in the surface layer, and 14 in the ground layer. Tussock-forming graminoids Molinia caerulea, Carex davalliana, and Sesleria caerulea dominated. Moss coverage (13 [+ or -] 7%) was much lower than in clusters 1 (28 [+ or -] 18%) and 2 (37 [+ or -] 20). In the ground layer Calliergonella cuspidata, Ctenidium molluscum, Drepanocladus cossonii, and Fissidens adianthoides were most frequent but with a low coverage. At the end of October the water table was close to the surface in depressions between tussocks but the seasonal amplitude of fluctuation was 4-25 cm.

Cluster 4, the Molinia caerulea Schoenus ferrugineus, was the most widely distributed group (releves 5, 6, 8, 9, 10, 14, 17, 22), which occupied the central and western parts of the fen site. We found 9 species in the shrub layer, 26 in the surface layer, and 19 species of bryophytes in the ground layer. In contrast to clusters 2 and 3, the coverage of shrubs (Betula humilis) was high in this cluster and dwarf shrubs (Andromeda polifolia, Empetrum nigrum) were present. On average the coverage of Molinia caerulea increased up to 24%. Nevertheless, the moss species composition reflected the calcareous-rich fen situation, but the distribution of most species was sporadic, with the mean coverage below 1%.

Only the coverage of Ctenidium molluscum was about 4% on average. Selaginella selaginoides was present in seven of the eight releves and Pinguicula alpina in one releve. In October 2007 the water level was at or up to 3 cm below the surface but seasonal fluctuations were notable (16-38 cm).

Cluster 5, the Molinia caerulea group, was distributed in the southern part of the area (releves 7, 15, 21, 23). The tree layer was formed by three species (Betula pubescens, Pinus sylvestris, and Populus tremula). The luxurious shrub layer of the cluster consisted of 8 species. Herbaceous plants (a total of 16 species found) were mostly tussock-forming graminoids, among which Molinia caerulea dominated with the coverage of up to 65% (mean 37%) and up to 40 cm (mean 25 cm) high tussocks. Dwarf shrubs (Ledum palustre, Vaccinium uliginosum, V. vitis-idaea) occupied stump hummocks together with certain forest mosses (Pleurozium schreberi, Rhizomnium punctatum, Rhytidiadelphus triquetrus) and peat mosses (mostly Sphagnum contortum, S. fallax, S. papillosum, S. warnstorfii). Other fen bryophyte species were present very sparsely. We found a total of 24 moss species in the ground layer. The releves of the cluster contained 12 moss species not present in other releves. At the end of October the water table was close to the surface, the summer minimum was 22-28 cm below the surface and seasonal fluctuation some 17.25 cm.

Environmental Conditions

At the end of October 2007 the water level between tussocks was close to the surface and varied rather little between plots. In the next spring (30 May 2008) and summer (30 July 2008) the DWL was from +3 to -24 cm and from -1 to -39 cm, respectively. In the northern, eastern, and central parts of the fen site the water table lay closer to the surface than around plots of clusters 4 and 5. The amplitude of the water level seasonal fluctuation (WLF) in plots between the three measurements varied from 1 to 38 cm. Pore water EC varied considerably between plots (from 220 to 840 [micro]S [cm^{-1}]), whereas only slight differences were observed in the values measured in late autumn and in midsummer.

Clusters 1, 2, and 3 had higher average EC values than clusters 4 and 5 (over 600 and less than 500 [micro]S [cm^{-1}], respectively). The pH of mire water varied only a little, from 6.1 to 7.1 between plots and between two years, 2007 and 2008. The coverage and height of tussocks formed by Carex davalliana, Schoenus ferrugineus (clusters 1, 2 and 3), or Molinia caerulea (clusters 4 and 5) varied from 15% to 60% and from 10 to 30 cm, respectively, between the plots. The coverage of Molinia caerulea was between 0.1% and 55%. The coverage of tussocks, their mean height, coverage of Molinia, depth to water table in July, and WLF between autumn maximum and summer minimum values were significantly interrelated. The DWL autumn maximum was not related to the DWL measured in May and July or to other environmental variables. The spring and summer values of the DWL and water EC of the samples collected in October 2007 were significantly correlated. The coverage of litter had no significant correlation with other environmental variables measured.

Ordination of Vegetation Analyses

We used the Mantel test to measure the relationship between the field and surface layer species compositions. No significant overall relationship was established between vascular plant and moss species coverage ($r = 0.09$ at $p = 0.18$ with $t = 1.33$) for the 23 plots. Therefore we applied NMS ordination separately for plant species from the field and the surface layer.

Vascular Plant Species

The ordination of 23 plots was achieved with a final stress of 13.64 and final instability of 0.014 after 25 iterations for the three-dimensional solution. The Monte Carlo test with 249 runs demonstrated that the final stress was significantly lower ($p = 0.004$) than that reached by chance. The proportion of variation, after rotation of the diagram by + 10°, explained by axes 1, 2, and 3 was 19.7%, 27.7%, and 32.4%, respectively (total for three axes 79.8%). The first ordination axis was correlated significantly ($p < 0.05$) with the pore water pH measured in October ($r = -0.49$) and with the EC measured in June ($r = -0.35$). The second ordination axis did not correlate significantly with any variable considered. The third ordination axis correlated significantly with the EC measured in October 2007 ($r = -0.629$) and in June 2008 ($r = -0.458$) and with WLF ($r = 0.577$). Axes 1 and 3 explained approximately 51.9% of the variation on the ordination space. We also found a relationship between the field layer species composition and the four environmental variables.

The standardised Mantel statistic r indicated a weak ($r = 0.19$) but significant (at $p = 0.034$) positive ($t = 2.12$) association between the two dissimilarity matrixes. The highest value of the standardised Mantel statistic was achieved when the EC in October and the WLF were fitted into the second matrix ($r = 0.23$, $p = 0.007$, $t = 2.70$). Clusters arrayed along the WLF-pore water EC gradient on axis 3. Virtually all plots from clusters 4 and 5 were widely distributed on the upper left and right quadrants of the ordination diagram. Plots from clusters 2 and 3 were distributed on the lower left quadrant, characterised by unimportant seasonal fluctuations of the water level and high pore water pH and EC values. The two plots belonging to cluster 1 were located on the lower right quadrant with a high pore water EC and stable water level. Axis 3 of the NMS ordination separated a group of vascular plant species found in the part of the fen site where the DWL was the deepest and the pore water EC values were considerably low. The other separate group of species was located in the lower right quadrant. These species were frequent in the part of the site with a stable water level lying close to the surface and with high EC values. The species located near the centre, such as Schoenus ferrugineus, Gymnadenia conopsea, Epipactis palustris, Saussurea esthonica, Primula farinose, and Selaginella selaginoides, were well correlated with axis 2, but nonsignificantly related with any parameter considered.

Bryophytes

The ordination of 23 plots was achieved with the final stress of 15.4 and final instability of 0.0076 after 35 iterations for the three-dimensional solution. The Monte Carlo test with 249 runs demonstrated that the final stress was significantly lower ($p = 0.004$) than that reached by chance. The proportion of variation explained by axes 1, 2, and 3 was 11.1%, 15.1%, and 17.3%, respectively (total for three axes 43.5%).

The first ordination axis correlated significantly ($p < 0.05$) with the pore water EC in October ($r = -0.61$) and with the DWL in July ($r = 0.60$). The third ordination axis correlated significantly with the coverage of tussocks ($r = 0.528$) and the mean height of tussocks ($r = 0.652$). The second ordination axis correlated significantly with the four variables considered and was well related to axes 1 and 3. Axes 1 and 3 explained about 28% of the variation on the ordination space. The standardised Mantel statistic r indicated that the dissimilarity matrixes of moss species and the four environmental variables were significantly (at $p = 0.034$) and positively ($t = 2.12$) associated ($r = 0.26$).

The species located in the lower left quadrant of the ordination diagram preferred a habitat with few low tussocks close to the summer level of surface water. The pore water EC had high values. The upper right quadrant close to axis 3 represents a group of moss species from plots with a high mean height of tussocks. Three species in the same quadrant (Dicranum bonjeanii, Lophocolea heterophylla, Bryum sp.) were found to occupy tussocks in the driest part of the fen site. The species in the upper left quadrant preferred near-surface midsummer water level with high EC values of pore water. The lower right quadrant contains no moss species located around the three plots (9, 10, and 17) that belonged to cluster 4. These plots were very poor in moss species and the number of individuals was low.

Habitat Conditions

Hydrologically the Paraspollu fen site includes sections of two types. In the northern and eastern sections the water level is high and stable (around clusters 1, 2, and partly 3). This phenomenon can be explained by the stable source of water inflow a spring on the eastern slope of the mire depression and by recharging groundwater. This part of the fen site showed a significant negative correlation between the amplitude of WLF and EC of pore water.

The southern and western sections of the site belong to the discharge zone as the water table drops down to about 40 cm in summer and shows great seasonal fluctuations. The EC values of pore water were quite low and more variable between the plots than in the plots of the recharge zone, which was possibly caused by the dilution effect of pure rainwaters. The pore water pH varied insignificantly from 6 to 7 units between plots and had somewhat higher values at the beginning of summer than in late autumn. One of the causes of the spatial and temporal stability of the pore water pH is the buffering effect of bicarbonates.

The Density and Coverage of Calcicole Species in Relation to Habitat Conditions

The proportions of calcicole vascular plant and bryophyte species were not related to the number of plots per cluster. The inter-plot differences in species diversity were most strongly related to differences in habitat conditions. The number of calcicole moss species was higher in plots of clusters 2 and 3 than in other clusters.

The coverage of calcicole vascular plant and moss species differed largely between the clusters. The much higher coverage of calcicole moss species in plots of cluster 2 than in other parts of the fen may be due to the stable and near-surface calcium-rich waters (as reflected by high EC values) and to the relatively low coverage of the field layer. The notably lower coverage of calcicole vascular plant species in cluster 5 and to a lesser extent in cluster 4 compared to clusters 2 and 3 is most probably connected with a significantly higher coverage of Molinia caerulea in the former. The expansion of M. caerulea in turn is caused by increasing DWL during the vegetation period. The optimum conditions for most of the calcicole vascular plant species represented in the Paraspollu fen site are the seasonal fluctuation of the water table by less than 20-25 cm and the midsummer water level not lower than 15 cm below the surface. Unlike vascular plants, calcicole moss species depend more on the water chemistry than on the substrate chemistry. Calcicole bryophytes probably require even a more stable and closer to the surface water level (9-10 cm and 4-5 cm, respectively, on the study site).

The coverage of litter varied largely between the plots (from 15% to 80%). Surprisingly, the effect of litter on the species composition and coverage of both vascular plants and bryophytes was insignificant. Hajkova & Hajek (2003) confirmed that high quantities of litter cause a decrease in bryophyte biomass but not in species richness. Peintinger & Bergamini (2006) found that in Swiss calcareous fens litter mass explains differences in the species composition of vascular plants equally well with the effect of abandonment, whereas the influence of litter on bryophytes is much less important. The experiment conducted in a Minnesota fen (USA) demonstrated that litter addition has a minor effect on the composition and production of plant cover species. Instead, litter removal, mainly because of increased light availability and soil temperature, leads to changes in the plant community in the fen. Our study site was abandoned at least some 40.50 years ago. Therefore, we may believe that such a time interval was sufficiently long for the system to reach a stable state between the rate of litter accumulation, plant species composition, and vegetation coverage.

Environmental Effect on the Distribution of Plant Groups

The Paraspollu fen site is characterised by a high variability in species composition and vegetation structure, depth of the water level, seasonal amplitude of water level fluctuations, and water conductivity gradients. About 52% of the vascular plant species variance was due to four environmental variables amplitude of WLF, water pH, and conductivity measured in late autumn and early summer. More than 28% of the variance can be explained by other habitat conditions, not measured by us, possibly by substrate chemistry (e.g. phosphorus availability).

It is often reported that the compositional pattern of the plant cover in fens is most importantly controlled by the water level and its stability or by the length of the period when the water level drops in summer. On a small scale (across a mire) also the soil organic matter content is of importance. For bryophytes, the surface microtopography has equally important control with hydrology and water conductivity, as the height and coverage of tussocks, EC of pore water, and midsummer minimum DWL describe some 43.5% of the compositional variation of moss species. Where the tussocks occupied about half of the surface and leaves of sedges and grasses shaded almost totally the intertussock small depressions (clusters 4 and 5), the shade-intolerant moss species were able to inhabit only the tops and sides of tussocks. In such a situation the effect of the midsummer minimum DWL was not as great as in plots of clusters 2 and 3 where the water level was at almost the same depth but differences in vascular plant cover were significant.

The absence of a relationship in the species compositional variation between vascular plants and bryophytes can be explained by the more rapid response of moss species to changes in the habitat conditions compared to vascular species. Clusters 1, 2, and 3 represent the near-natural calcareous-rich fen state. The Carex panicea Schoenus ferrugineus Drepanocladus cossonii assemblage of cluster 2 corresponds well to the calcareous tufa-forming fen type according to Hajek et al. (2006). In small depressions between tussocks calcium carbonate precipitates and forms unconsolidated marl. The number and coverage of calcareous species, especially moss species, was higher in the Carex panicea. Schoenus ferrugineus Drepanocladus cossonii assemblage compared to the other assemblages distinguished on the Paraspollu fen. In the moss cover the typically calcicole species D. cossonii dominated with Campylium stellatum as a codominant.

The Phragmites australis Calliergonella cuspidata assemblage (cluster 2) is distributed in almost the same hydrological conditions as the Carex panicea. Schoenus ferrugineus Drepanocladus cossonii assemblage. Phragmites australis forms a rather dense cover and calcium does not precipitate, most likely because of mobile but stable near-surface groundwater. Moss layer dominants Calliergonella cuspidata and Campylium elodes indicate calcareous and extremely rich fen according to Hajek et al. (2006). The Molinia caerulea Carex davalliana assemblage (cluster 3) can also be classified as calcareous and extremely rich fen type but weakly affected by drainage.

Two assemblages, namely Molinia caerulea Schoenus ferrugineus (cluster 4) and Molinia caerulea (cluster 5), demonstrate that even moderate drainage promotes the expansion of M. caerulea. The principal difference between these assemblages is that S. ferrugineus was frequent in the former but occasional in the latter assemblage. In the Molinia caerulea assemblage certain forest floor species such as Hylocomnium splendes and Pleurozium schreberi dominated among the bryophytes and several Sphagnum species (S. contortum, S. fallax, S. papillosum, S. warnstorfii) had formed small patches on stump hummocks. The long-term (over 20 years) monitoring of the drainage effect on the vegetation of a rich fen in Central Sweden (Malson et al., 2008) showed that rapid emergence of a few dominants such as M. caerulea, Betula pubescens, and Sphagnum spp. may start at different times after drainage. The Molinia cover made up a few per cent just after drainage in 1979, but over 50% 10-30 years later.

Distribution of Rare Species

The rarest species among vascular plants on the Paraspollu fen site were Malaxis monophyllos, Selaginella selaginoides, and Pinguicula alpina, which are listed in category II of protected species in Estonia. Malaxis monophyllos is a sporadic species growing in moist or flooded sites of fens and mire forests. In Estonia its distribution has decreased due to extensive drainage activities. On the Paraspollu fen this species was found in one releve of cluster 2 in a wet habitat with negligible seasonal fluctuations of the water table around the ground surface.

Selaginella selaginoides is a relict of the post-glacial period, which is on the southern border of its distribution area in Estonia. On our study site S. selaginoides was closely related to the tussock-forming graminoid Schoenus ferrugineus growing on the top and sides of its tussocks. Pinguicula alpina, a very rare species in Estonia, grows in calcareous fens. On the Paraspollu fen it was found in one releve in cluster 4 on the sides of S. ferrugineus and Carex davalliana tussocks. Our data are consistent with the results of Trass (1986) indicating that Epipactis palustris gains from weak drainage. Saussurea alpina subsp. esthonica also seems to benefit from weak drainage and increasing shade provided by a sparse tree and shrub layer. The orchid species present on the Paraspollu fen site prefer wet areas or those moderately affected by drainage. The water level there was not deeper than 20-25 cm during the midsummer minimum.

The bryophyte list of the studied fen contains several rare and protected species preferring calcareous substrate, such as Campylium calcareum, C. halleri, and Catoscopium nigritum. Only the last species was identified in wetter conditions, whereas the two Campylium species with a very low coverage were found in the driest, Molinia-rich part of the fen site. We distinguished five assemblages on the Paraspollu fen well arranged on the gradient of the midsummer water level minimum together with the respective late autumn pore water conductivity pattern. The proportion of calcareous moss species decreased rapidly when the seasonal amplitude of the water level fluctuation reached 9-10 cm and midsummer water level minimum was more than 4-5 cm. Calcareous vascular plant species were more tolerant to water level instability and still dominated in plots where the water level fluctuated with an amplitude of up to 25 cm.

The microtopography had also a strong effect on the coverage and species composition of mosses. Obviously different responses to hydrology and microtopography are among the main reasons for the weak (statistically nonsignificant) correspondence between the compositional variance of field and surface layer plant species on the Paraspollu calcareous fen site. The fen site, weakly drained in part, is a habitat of many rare and endangered vascular and moss species. The drainage of the abandoned calcareous fen site favours the expansion of Molinia caerulea and encroachment of woody species. There is an urgent need to develop a well-defined management plan to restore the calcareous rich fen plant cover and create respective habitat conditions over the site based on good knowledge of nutritional requirements of plant species, interspecies relationships, and substrate biogeochemistry.

Chapter 7

The Molecular Basis

Tolerance to a drug, including alcohol, was first described as a form of behavioural plasticity and was defined as a decreased response to repeated drug exposures. Tolerance can be described at different levels of biological complexity molecular, cellular, and behavioural or by its temporal characteristics of how rapidly the alcohol affects the organism. Behavioural tolerance is measured at the level of the activity of an entire animal, resulting from mutual interactions in several brain structures and with other systems. For example, a simple walk on a straight wooden beam is a good measure of a mouse's coordination and sense of balance. Several brain regions (e.g., visual cortex, occipital cortex, cerebellum) "talk" to each other, as well as with other systems (e.g., muscular-skeletal) to enable the mouse to successfully walk from one end of the beam to the other. Alcohol intoxication impairs this complex coordination and a "drunken" mouse falls off the beam. An alcohol-tolerant mouse can, at least partially, retain its coordination and walk on the beam almost as well as an alcohol-abstinent mouse, in spite of the ingestion of alcohol. Previous studies by Hoffman and Tabakoff (1989) have stressed the difference between "intrinsic" tolerance, which results from changes within the neurons controlling a bahaviour, and "extrinsic" tolerance, which results in behavioural adaptation via alterations in compensatory neural circuits (i.e., not via molecular tolerance within the primary neural circuit).

Currently, behavioural tolerance is divided into three categories: acute, rapid, and chronic. Acute tolerance develops within a single drinking session, typically within minutes, whereas chronic tolerance occurs after a longer time, usually following days of continuous or intermittent alcohol exposure. Rapid tolerance shares many similarities with chronic tolerance but develops faster, typically within 8 to 24 hours. Cellular tolerance typically is assessed at the level of a neuronal tissue consisting of a network of many neuronal and supportive cells, or even a single neuron. Knowledge of molecular tolerance comes from dissecting adaptational processes developed by individual molecules (e.g., ion channels) during exposure to a drug. The current notion is that even complex behavioural traits can be traced to individual molecules. All of alcohol's complex effects on an organism start at the molecular level, during interaction of an alcohol molecule with its molecular target(s).

Although these interactions at the molecular level are complex, the level of complexity increases at the cellular and behavioural level because of multiple, complex, intertwined pathways that contribute to the development of tolerance. It still is unclear how interactions with molecules in the brain eventually can lead to the altered bahaviour defined as alcohol dependence. However, there is a great deal of information on the mechanism of tolerance at the molecular level; thus, this review will focus primarily on that level and discuss potential relationships with other levels, such as acute, rapid, and chronic tolerance at the behavioural level, when possible.

At the molecular level, our discussion will first address mechanisms involving modification of a mature, functional ion channel present in the neuronal plasma membrane and then move progressively upstream to the nucleus, site of the "birth" of the channel protein. The approach in this chapter will be to focus on a single channel type, the BK channel a potassium ion channel of high conductance regulated by both voltage and calcium that has been implicated in the onset of tolerance. This review focuses on the BK channel for several reasons: (1) This channel is abundantly expressed in many brain regions; (2) it plays an essential role in many aspects of neuronal physiology, including neurotransmitter release, shaping of action potentials, and dendritic integration; (3) its activity is regulated by alcohol; (4) it is one of the key elements of behavioural tolerance to alcohol (at least in invertebrates); and (5) it is the model that the authors are most familiar with. However, whereas the focus on a single channel protein is useful, it means that significant work by many laboratories examining other alcohol targets, which might well prove equally important, is not extensively covered. When applicable, other molecules relevant to the development of alcohol tolerance are noted.

Molecular Basis of Tolerance

Alcohol can regulate a membrane-bound ion channel in several ways: by changing the activity of proteins through posttranslational modifications, interacting with membrane lipids, interacting with auxiliary proteins, modulation of membrane protein expression (i.e., trafficking), and altering the spatial organisation of membrane proteins. Alcohol's effects on processes that regulate channel protein production, such as mRNA expression in the cytoplasm and gene expression in the nucleus, also play a role in the development of tolerance. All of these molecular processes act in concert to contribute to the development of tolerance. Nevertheless, tolerance to alcohol has been examined in a limited number of behavioural parameters, and it is clear that the relationship between molecular mechanisms of tolerance and behavioural responses is far from understood. The following sections examine alcohol's effects on these molecular mechanisms in greater detail.

Effects on the Activity of Proteins

Posttranslational Modifications

Alcohol begins to increase activity of the BK channel within seconds of exposure. Current thinking is that this activation is unlikely to be mediated by second messenger' systems but rather results from a direct interaction of alcohol with the channel, leading to modification of channel gating, which regulates the ratio between channel open states (during which ions flow through a channel) and closed states (during which ions cannot flow through a channel). Specifically, alcohol increases the contribution of long openings and decreases the contribution of long closures, making the BK channel more active. Among the reasons for thinking that a direct interaction is likely are findings from studies in which the BK channel is isolated from its host membrane and incorporated into an artificial bilayer. When single-channel recordings then are made from this isolated channel, results indicate that alcohol is capable of modifying the open probability (an electrophysiological measurement indicating the proportion of the total recording period that a channel has been open, allowing ions to flow through it). This result indicates that alcohol acts directly on the isolated channel, increasing its potential for remaining open in a manner similar to that seen in the native membrane. However, it must be noted that an alcohol-binding site on the BK channel has not yet been identified. It is conceivable that proteins such as kinases and phosphatases may be carried along with the incorporated channel and that the direct action of alcohol, even in this highly reduced preparation, reflects the actions on those associated proteins. However, it also can be argued that, functionally, these enzymes are part of the "channel" complex in terms of discussing the direct effect of alcohol.

Often, this potentiation of channel opening is transient and subsides after several minutes, thus exhibiting acute tolerance. This type of tolerance can be controlled by changes in the channel's phosphorylation state. The phosphorylation state of any protein (including ion channels) reflects the balance between the actions of protein kinases, which transfer a phosphate group from adenosine triphosphate to the protein (i.e., phosphorylation), and protein phosphatases, which remove that group (i.e., dephosphorylation). Our recent data indicate that kinases and phosphatases are involved in alcohol potentiation of the BK channel and the development of tolerance. For example, although BK channels seem not to be constitutively phosphorylated by protein kinase A (PKA), phosphorylation by this kinase is necessary for alcohol potentiation of the BK channel. Interestingly, the additional PKA phosphorylation site present in an isoform protein of the BK channel, encoded by the alternatively spliced exon STREX, prevents potentiation by alcohol. Clearly, the phosphorylation state of the BK channel is important for its sensitivity and the development of acute tolerance. Alcohol also regulates phosphorylation of other important membrane proteins, such as receptors for [*c*]-aminobutyric acid ([$GABA_A$]) and N-methyl-D-aspartic acid (NMDA). As described more extensively in recent reviews, alcohol-induced alterations in the phosphorylation state of these receptors contribute to the development of tolerance.

Collectively, posttranslational modifications of ion channel proteins contribute to the development of tolerance to alcohol. Ion channels, functioning in the neuronal plasma membrane, undergo phosphorylation (and dephosphorylation), which modifies their activities. Alcohol can affect this phosphorylation state and create a new phosphorylation set point, characteristic of alcohol-tolerant channels.

Interaction With Membrane Lipids

In addition to alterations of the protein, such as the addition or removal of phosphate groups, more indirect effects, such as a change in the immediate lipid microenvironment, can alter alcohol pharmacology. The plasma membrane is composed of a heterogeneous population of various lipids, such as sphingomyefins, phosphatidylocholines, and cholesterol, which determine lipid bilayer thickness and rigidity. Lipid structures, which can act to segregate proteins, are microdomains called rafts. For many years, membrane lipids were thought to be the primary site of alcohol action. More recently, however, largely because mutagenesis studies indicate that very small changes in secondary structure can lead to very large changes in alcohol sensitivity, the main focus has switched to alcohol's direct interaction with membrane proteins.

However, emerging evidence indicates that lipids play a significant role in the response of imbedded proteins to alcohol and the development of alcohol tolerance. Altering the thickness of the lipid bilayer affects the time course of the acute response to alcohol and can completely prevent alcohol potentiation of the BK channel. BK channels embedded in a thin bilayer are strongly potentiated by alcohol, whereas channels present in a thick bilayer are inhibited. Moreover, cholesterol, a main component of lipid rafts, regulates BK channel sensitivity to alcohol. Increased cholesterol in the plasma membrane can diminish the alcohol potentiation of the BK channel, potentially contributing to channel tolerance. Thus, the lipid microenvironment plays an essential role in the development of acute tolerance. Both acute and chronic alcohol exposure can change lipid content particularly that of cholesterol in cells. Short alcohol exposure depletes cholesterol, presumably decreasing also the amount of plasma membrane bound cholesterol, making membranes more fluid or disordered. In contrast, after days-long exposure to alcohol, cellular membranes become tolerant to the disordering effects of alcohol (so-called "membrane tolerance, through alterations in specific phospholipids (cardiolipin, phosphatidylinositol) and increased cholesterol content.

How, exactly, membrane tolerance might contribute to behavioural tolerance is not yet clear. Nevertheless, it is possible that the altered amount of cholesterol in a microdomain surrounding the BK channel caused by either acute or chronic alcohol exposure can temper alcohol's effect on BK channels and thus contribute to the development of acute or chronic tolerance. In summary, alcohol exposure changes the lipid composition of the plasma membrane. The new lipid environment can affect ion channel sensitivity to alcohol and contribute to either acute or chronic tolerance, or both.

Interaction With Auxiliary Proteins

BK channels, like many other membrane proteins, are clustered with other (auxiliary) proteins within lipid rafts. BK channels also can form functional dyads with voltage-dependent calcium channels and bind to a multifunctional scaffolding protein known as [RACK1], or [β]2 adrenergic receptors. Auxiliary [β]-subunits are tightly bound proteins that affect BK channel characteristics such as its sensitivity to calcium, an important molecule in cell-to-cell communication. Recently, it was discovered that [β]-subunits of the BK channel also can control the alcohol response of these channels and influence the development of tolerance. For example, upon cotransfection of HEK293 cells with both the BK channel main a-subunit and an auxiliary [β]1-subunit, the resulting BK channel is composed of [α]-[β]1 complexes. The association of the [β]1-subunit with the [α]-subunit prevents the potentiation of the channel by alcohol. Interestingly, the [β]4-subunit also reduces this potentiation. The molecular mechanism of [β]-subunit regulation of BK channel alcohol sensitivity remains unclear. Expression of BK [β]-subunits varies among brain regions. BK [β]4-subunits are abundant throughout the brain, whereas expression of BK [β]1-subunits is, in general, very low.

Recent studies indicate that the BK [β]4-subunit exerts a profound influence on the development of acute tolerance. BK channels from wild-type mice show little alcohol tolerance compared with channels from BK [β]4-knockout mice, which exhibit acute tolerance. Moreover, BK [β]4's role in the development of tolerance also can be observed on a level of neuronal activity. BK channels contribute to the frequency of electrical signals (i.e., action potentials, also known as spikes) generated by neurons. Alcohol persistently decreases the number of action potentials in striatal slices harvested from wild-type mice. However, the change in action potential frequency lasts only a few minutes in the same preparation harvested from BK [β]4-knockout mice. This transient change in spike patterning is consistent with the development of tolerance. [β]4-knockout mice also develop behavioural rapid tolerance to the locomotor effects of alcohol, in contrast to the wild-type mice. Most striking, the [β]4-knockout mice consume significantly greater amounts of alcohol, consistent with human studies, which indicates a correlation between the degree of acute behavioural tolerance and the likelihood of future alcohol abuse.

Changes in subunit composition as a potential mechanism in the development of tolerance is not limited to the BK channel. It is well established that alcohol selectively alters mRNA and protein expression of selected subunits of membrane proteins such as NMDA and GABA receptors. For example, chronic alcohol exposure selectively increases expression of the NMDA receptor R2B subunit protein and subsequently increases its surface expression and targeting to synapses. This mechanism contributes to the alcohol-triggered upregulation of NMDA receptor function and the development of alcohol tolerance. Similarly, chronic alcohol exposure results in the reorganisation of [$GABA_A$] receptor subunits, such as the downregulation of [$GABA_A$] receptor and [α]1-subunits or the down- or upregulation of the [α]4-subunit (depending on its neuronal location). These changes, in turn, contribute to [$GABA_A$] receptor-related alcohol tolerance.

From the allostatic point of view, modulation of specific subunits of BK, NMDA, and [$GABA_A$] receptors (and probably other channels and receptors as well) establishes new expression levels of these subunit proteins. To maintain channel function, new "equilibria" of subunits within a channel are formed, allowing "stability through change." Such modified channel subunit composition contributes to the development of tolerance.

Can it be stated that allostatic changes in subunit composition of ion channels contribute selectively to acute, rapid, or chronic types of tolerance? Most experiments to date have been performed using alcohol exposures lasting several days, thus relating change in subunit composition to chronic tolerance. However, there is some evidence indicating that subunit reorganisation can be rapid. Recent data by Liang and colleagues (2007) demonstrate that as early as 1 hour after alcohol exposure, cell surface expression of [$GABA_A$] receptor auxiliary subunits a4 and (but not others) is significantly decreased. These results suggest that alcohol-triggered reorganisation of channel subunits also could contribute to acute tolerance. Further studies determining the role of auxiliary proteins in other types of tolerance are warranted.

Modulation of Membrane Protein Expression

Alcohol's modulation of channel cell surface expression is another molecular mechanism contributing to the development of tolerance. In neurons, the surface expression of functional membrane proteins determines neuronal properties and is highly regulated through several mechanisms. The density of channels on the neuronal surface is a precisely controlled balance among delivery of newly produced channels, recycling, internalisation, and degradation.

Similar alcohol regulation of the internalisation of [$GABA_A$] receptors in neurons has been reported. Chronic alcohol exposure causes increased internalisation of [$GABA_A$] receptors. To fully understand the contribution of altered trafficking of channels and receptors to a particular class (e.g., acute, chronic) of tolerance, it will be essential to determine whether these alterations are protein synthesis dependent. Some hints are coming from a recent BK channel gene regulation study indicating that alcohol invokes changes in BK mRNA level, which precede altered trafficking of the channel.

Spatial Organisation of Membrane Proteins

Some channel types (such as BK) are not distributed homogeneously on the neuronal surface but occur in smaller clusters of BK channels, or microdomains. It has been observed that alcohol alters the cell surface expression of BK channels in neurons, similar to a mechanism of tolerance reported for other drugs. In presynaptic terminals, where BK channels are expressed in clusters, alcohol causes declustering of BK channels in the plasma membrane and redistribution of channels between the membrane and cytoplasm (observed as decreased channel density in the plasma membrane and increased density of channels in the terminal interior). These morphological changes are concurrent with a loss of BK alcohol sensitivity and/or tolerance. Interestingly, chronic alcohol exposure also can enhance clustering of some other membrane proteins. The expression of NMDA receptor type 1 and 2B clusters is increased on the surface of cortical neurons, as well as hippocampal neurons, specifically in synaptic boutons.

Spatial organisation of membrane proteins (clustering) is mediated by scaffolding proteins, which link them to specific signalling cascades and other cytoskeletal elements. Scaffolding proteins also are involved in the development of alcohol tolerance. For example, it has been shown that the scaffolding protein [RACK1] organises NMDA receptor complexes and plays an important role in the development of acute alcohol tolerance of these receptors. Acute alcohol exposure inhibits NMDA receptor activity. Additionally, alcohol disrupts NMDA receptor binding to [RACK1], allowing for subsequent phosphorylation of the NR2B subunit of the NMDA receptor by Fyn kinase. This phosphorylation gradually reduces alcohol- induced inhibition of the NMDA receptor, thus contributing to the development of acute tolerance.

Homer, another neuronal scaffolding protein recently implicated in the development of alcohol tolerance, has been shown to play a role in synapse formation, neuronal excitability, and addiction-related neuroplasticity. Homer proteins form tetramers, in which four products of Homer mRNA (each called a monomer) create a single functional structure. Each Homer tetramer provides a cytoplasmic framework for receptors and channels on the plasma membrane and links them to important components inside the cell, such as intracellular messenger systems and other scaffolding proteins. They also are part of the NMDA receptor complex, and deletion of Home2 decreases the amount of NMDA NR2B subunit on the cell surface.

The role of Homer in alcohol tolerance has been suggested by recent fruit fly (D. melanogaster) studies. Mutant flies without functional Homer do not develop behavioural rapid tolerance, indicating an essential role for this protein in this particular type of tolerance. Moreover, rapid tolerance is "rescued" (meaning the ability to develop tolerance is maintained) by expression of wild-type Homer in neurons, specifically in a subset of neurons in the ellipsoid body located in the head, which are involved in higher control of locomotor activity such as walking and leg coordination. The role of Homer in other classes of tolerance has not yet been determined. An especially important research goal would be to determine whether similar mechanisms might be observed also in the mammalian system.

[RACK1] and Homer proteins are scaffolding elements and, as such, are tightly bound to the actin cytoskeleton. Actin fibres form a complex meshwork in the cell and constantly and quickly are remodelled in response to internal and external cues. Altered actin dynamics recently have been linked to alcohol tolerance. In wild-type mice, acute alcohol exposure causes significant loss of filamentous actin in neurons, presumably by targeting epidermal growth factor receptor pathway substrate (Eps8), a member of the protein family regulating actin polymerisation. This alcohol-related loss of actin can be reversed quickly by washing the drug away. In animals lacking Eps8, actin stability is enhanced, causing resistance to the destabilising effects of alcohol. Additionally, Eps8 association with the NMDA receptor complex is thought to play a role in the development of molecular acute tolerance, as well as behavioural acute tolerance to the intoxicating effects of alcohol. An obvious question is how these effects on actin contribute to rapid or chronic tolerance in mice.

To date, molecular mechanisms regulating protein cell surface expression seem to be involved in the development of specific classes of alcohol tolerance (i.e., rapid tolerance is linked to the Homer protein;

acute tolerance to actin and [RACK1]). More research is needed to determine whether these molecules also contribute to other classes of tolerance and if they are molecular markers for specific classes of tolerance.

Effects on Protein Dynamics

Epigenetics

Epigenetics refers to heritable changes in gene expression without alterations in the DNA sequence. In addition to alcohol's more direct effects on membrane and associated proteins, it also can affect upstream processes controlling channel production (namely, transcription and translation), including epigenetic influences on gene expression. Several molecular mechanisms of epigenetic regulation of gene expression exist. Some epigenetic mechanisms take place in the nucleus (DNA methylation, posttranscriptional modifications of histones, chromatin remodelling), whereas others occur in the cytoplasm (mRNA processing). There is growing evidence that at least some of these epigenetic mechanisms are involved in the development of tolerance, as described below.

Epigenetic Cytoplasmic Mechanisms

Alcohol is known to modify mRNA expression of many different proteins, including the BK channel pore-forming [α]-subunit. Typically, the a-subunit of the BK channel undergoes alternative splicing, giving rise to a number of final transcripts varying in their exonal structure. Researchers recently have discovered that different BK channel transcripts encode channels with different alcohol sensitivities. BK channels containing an insert encoded by an alternatively spliced exon called ALCOREX are highly potentiated by alcohol, whereas others containing an insert encoded by an alternatively spliced exon called STREX are insensitive to alcohol. Pietrykowski and colleagues (2008) found that alcohol causes remodelling of the BK channel transcript landscape by regulating the stability of a subset of these transcripts, such that the relative number of transcripts encoding highly sensitive channels are decreased, whereas that of transcripts encoding less sensitive channel proteins are increased, consistent with tolerance. Differential regulation of transcript stability occurs via a microRNA-based pathway.

MicroRNAs are recently discovered, small, noncoding RNA molecules regulating mRNA and protein expression. Alcohol exposure specifically upregulates one microRNA species, miR-9, which subsequently binds to selected BK mRNA transcripts containing a miR-9-binding site in their 3 -untranslated region (3 -UTR). This binding destabilises transcripts with miR-9 complementarity and causes their degradation. By this mechanism, the number of transcripts is reduced and their absolute and relative amounts are altered. The overall effect is a change in the contribution of each transcript product to BK channel assembly, causing an increase in the representation of alcohol-tolerant channels.

Thus, alcohol regulates gene expression using a microRNA-based mechanism. For the outcome of this regulation to be evident, new protein must be translated from modified transcripts, and delivered to its functional home. Because this process will take time, this mechanism is unlikely to contribute to acute tolerance and is more likely to play a role in rapid and chronic tolerance, which are both protein synthesis dependent in mammals. Contribution of this epigenetic mechanism to protein synthesis-independent types of tolerance, like acute or rapid tolerance in flies, is unlikely.

Alcohol regulation via a miR-9 dependent mechanism has the potential for a wide-ranging role in plasticity because many potential miR-9 targets also are regulated by alcohol. The role of other small, noncoding RNAs in alcohol tolerance has not yet been addressed. However, there is growing evidence that many other microRNA species also are regulated by alcohol. Thus, further studies are warranted to determine the important role of these newly appreciated regulatory molecules in the development of alcohol tolerance.

Epigenetic Nuclear Mechanisms

Epigenetic nuclear mechanisms also have been implicated recently in the development of alcohol tolerance, using a fruit fly homolog of the BK channel (Slowpoke, Slo) as a model. Gene transcription is regulated by chromatin structure via post-translational modifications (e.g., acetylation) of chromatin core proteins known as histories. Acetylation of histories weakens the association between histories and DNA, allowing binding of a transcriptosome complex to gene promoter regions, and subsequent transcription.

Brief exposure to alcohol causes increased Slo gene expression in the central nervous system of Drosophila. Enhanced expression of the Slo gene is mediated by increased histone acetylation within the Slo promoter region. Blocking of histone deacetylation by inhibition of histone deacetylase has a similar effect, confirming the involvement of this epigenetic pathway in the regulation of Slo gene expression. This mechanism exposes the Slo promoter to the CAMP response element-binding protein (CREB). CREB binding to the Slo promoter region enhances Slo gene transcription. Thus, activation of this pathway leads to changes in Slo protein expression, which contributes to the development of tolerance. Change in Slo expression can be seen as an attempt to maintain "stability through change" in accord with an allostatic mode of adaptation.

Genetics of Tolerance

Invertebrate models also are extensively used to determine links between genes and bahaviour, including the development of tolerance to alcohol. The Slo gene is one of the best examples of this link. It has been shown in both fruit fly and worm (C. elegans) that the behavioural response to alcohol is modified by targeting Slo channel function. In Drosophila, over expression of Slo in neurons produces a phenotype mimicking alcohol tolerance. A neuron-specific loss-of-function mutation of Slo produces fruit flies unable to develop rapid tolerance to alcohol. Similar observations have been made in worms in Slo's acute response to alcohol. These results suggest that the establishment of a new set point for Slo channel expression is important in the development of tolerance to alcohol.

Fruit fly studies provide evidence for the involvement of a number of molecules and systems in the development of tolerance. Fruit flies containing a loss-of-function mutation in the gene encoding tyramine [β]-hydroxylase are unable to synthesize octopamine (an invertebrate analogue of vertebrate's noradrenaline) and have a diminished ability to develop rapid tolerance to alcohol. Therefore, these results suggest the involvement of the noradrenergic system in the development of tolerance. Indeed, researchers have suggested a role for the adrenergic system in alcohol tolerance in mammals. In mice, the destruction of noradrenergic systems in brain by intraventricular injection of 6 hydroxydopamine prevents the development of chronic tolerance to alcohol. Interestingly, mutant flies with defective biosynthesis of octopamine display reductions in rapid tolerance only, whereas chronic tolerance is unchanged.

A newly described gene has been found to contribute to the development of rapid tolerance to alcohol in fruit flies. Hangover loss-of-function mutant flies show significantly reduced tolerance to alcohol compared with wildtype flies. Alcohol does not change Hangover expression. The exact role of the Hangover protein is unknown, although the predicted Hangover protein structure suggests its role in transcription. Thus, it is possible that alcohol, via the Hangover protein, can modulate transcription of other genes involved in the development of tolerance to alcohol. Hangover mutant flies also are defective in their response to environmental stress, supporting an important role for stress in the development of alcohol addiction and tolerance. Recent genetic studies suggest a role for Hangover in the development of alcoholism in humans. A single-nucleotide polymorphism in the human ortholog of Hangover (ZNF699) recently was correlated with alcohol dependence in Irish populations.

Other fruit fly screens point to additional genes and pathways that can contribute to the development of alcohol tolerance. Berger and colleagues (2008) screened 60 long-term memory mutants and discovered that several of these mutants display reduced rapid tolerance, others show decreased chronic tolerance, and one mutant demonstrates increased chronic tolerance.

New data also are coming from vertebrate studies providing insights into genetic influences on alcohol tolerance and also indicating that different types of tolerance can be genetically determined. Recent microarray analysis has identified at least eight candidate genes important for the development of acute tolerance to alcohol using inbred strains of mice. Moreover, another mouse study suggests that genetic factors influencing acute tolerance can play a major role in the development of rapid tolerance. Finally, successful breeding of mice displaying high and low rapid tolerance provided evidence for both a genetic contribution to rapid tolerance and for a genetic link between rapid and chronic tolerance. Further elucidation of these types of genetic influences underlying alcohol-relevant phenotypes may help to define the molecular elements of allostasis important in the development of different classes of tolerance.

Work from a number of laboratories has highlighted novel molecular mechanisms underlying the development of tolerance to alcohol. An emerging model indicates that multiple, complex, and intertwined molecular pathways contribute to the development of different classes of tolerance to alcohol. Tolerance typically is divided into three functional categories (acute, rapid, and chronic), but determining precisely which molecular underpinning underlies which class of tolerance (or if they are exclusive) can be difficult. There are multiple steps that can influence the response of surface proteins to alcohol, from gene transcription in the nucleus, to cytoplasmic regulation, to final expression in the plasma membrane. These steps are interlocked and sometimes difficult to separate and should be seen as a "continuum" rather than completely independent entities. As described here, all of these steps, including posttranslational modifications of proteins existing in the plasma membrane, their interactions with the lipid microenvironment, association with auxiliary proteins and cytoskeleton, trafficking, regulation of mRNA stability, and epigenetic and genetic mechanisms, contribute to the development of tolerance. Many more experiments need to be performed to determine the exact molecular mechanisms of each type of tolerance to alcohol and ultimately to define a molecular switch (or switches) allowing for transition from one type of tolerance to another. So far, one general conclusion seems to be that mechanisms which do not involve protein synthesis are related to acute tolerance in mammals and acute and rapid tolerance in fruit flies, whereas mechanisms upstream of protein synthesis are involved in rapid tolerance in mammals and in chronic tolerance in both mammals and fruit flies.

Importantly, each of these molecular mechanisms can be envisioned to contribute to a well-balanced set point according to allostatic principles. Alcohol exposure creates a new balance at each step, maintaining "stability through change." It is likely that alcohol can trigger these mechanisms simultaneously and that the central nervous system adapts to alcohol by dynamic interaction of these mechanisms. This integrated reaction provides the molecular bases for allo-static changes associated with the development of alcohol tolerance. The molecular underpinnings of tolerance described in this review, including membrane lipids, regulatory RNA molecules, and channel protein subunits, may provide potential therapeutic targets for alcoholism treatment. However, a rational approach to drug development will be facilitated by a clearer understanding of the relationship between molecular tolerance and the elements of addiction, such as dependency and craving, which have not been addressed in this review.

Chapter 8

Carbon Dioxide Concentrations Amplify Alternaria Alternata Sporulation

Anthropogenic increases in global atmospheric carbon dioxide ([CO_2]) concentration have been shown to stimulate earlier and greater production of allergenic pollen, as have warming temperatures. The effects of anthropogenic climate change on the production of airborne fungal spores are not as well documented, but the implications for allergic disease are no less important. As with pollen, exposure to fungal spores is associated with allergy and asthma symptoms, although the specifics of the relationship are not completely understood. Among patients with asthma from six regions of the world, 11.9%, on average, were sensitized to Alternaria alternata, with the proportion as high as 28.2% in Portland, Oregon; in addition, sensitivity to A. alternata was more prevalent among patients with more severe asthma. Allergenic fungal spores may be increasingly abundant in some areas of the globe as well. In Derby, United Kingdom, mean seasonal airborne spore concentrations of the genus Alternaria have increased over the years 1970-1998, as have the number of days with spore counts > 50 per cubic metre of air, and the start of the Alternaria spore season has advanced from the end to the beginning of June.

Rising atmospheric [CO_2] concentration is well documented [Intergovernmental Panel on Climate Change (IPCC) 2007]. Large increases in atmospheric [CO_2] concentration (e.g., doubled or greater increases in concentration) have been shown to alter both plant biomass and chemistry in many plant species. The responses of individual plant species to large increases in atmospheric [CO_2] concentration are variable but often include greater total biomass production and greater carbon-to-nitrogen ratios (C:N) of plant tissues. Primary consumers, including allergenic fungi that grow on living or dead plant materials, are therefore likely to encounter changes in substrate quality or quantity as global atmospheric [CO_2] concentration increases.

The responses of plant-fungal interactions to increased atmospheric [CO_2] concentration may be more complex and difficult to elucidate than the observed changes in pollen production and allergenicity. Fungi play several different roles in plant biology, acting as plant pathogens, saprobes (decomposers of dead plant material), or plant symbionts, with neutral, beneficial, or negative impacts on living plant hosts.

There may be a number of indirect effects on fungal growth and reproduction as a result of plant changes under increased atmospheric [CO_2] concentration as well as feedback and interactions among plants, fungi, and abiotic (e.g., environmental, nonorganismal) factors. In a field experiment using open-top chambers to grow individual poplar trees in ambient or doubled atmospheric [CO_2] concentration, Klironomos et al. (1997) found that air and leaf litter in chambers with doubled [CO_2] concentration contained sig-nificantly more fungal spores than those in ambient chambers.

An indirect effect of doubled atmospheric [CO_2] concentration on sporulation, mediated by changes in plant chemistry, was suggested by their findings; however, a direct effect of atmospheric [CO_2] concentration or of related changes such as soil moisture and relative humidity could not be ruled out.

To directly assess the quantitative and qualitative impacts of rising atmospheric [CO_2] concentration on allergenic fungi, we grew the perennial [C_3] monocot timothy grass (Phleum pratense), a common fodder used for hay, at four levels of atmospheric [CO_2] concentration approximating preindustrial, current, and potential/projected future concentrations (300, 400, 500, and 600 [micro]mol/mol, respectively). We used leaves from these plants to separately grow two fungal species. The first species, A. alternata, is a ubiquitous, facultatively plant-pathogenic or saprobic species known to produce allergenic airborne conidial spores.

The second species, Cladosporium phlei (syn. Heterosporium phlei), is a specific pathogen of timothy that has not been specifically shown to produce allergenic spores; other species within the genus Cladosporium are known to cause allergies in humans. We also examined the sporulation of A. alternata on clippings from field-grown grasses to see if responses were similar on other grass species. Our objective was to quantify changes in sporulation on plant leaves grown at varying atmospheric [CO_2] concentrations and elucidate the mechanism that drives such changes.

General Approach

We used two growth chambers to provide atmospheric [CO_2] concentration at 300, 400, 500, or 600 [micro]mol/mol to experimental timothy plants for 60 days of growth. These approximate atmospheric [CO_2] concentrations at the beginning of the 19th century, current ambient levels, and those projected by the years 2025 and 2040, respectively (IPCC 2007). During repeated runs over time, each of the two growth chambers was used for each level of [CO_2] concentration twice. During each run, each chamber contained 10 timothy plants. After 60 days of growth, leaf subsamples from each plant were used to grow A. alternata and C. phlei separately for 7 days inside loosely capped media bottles in an incubator. Media bottles were then volumetrically flooded with water, shaken, and subsampled for spore counts and, for A. alternata only, for antigenic protein extraction and quantification via ELISA using polyclonal antibodies.

Growth, harvest, and C:N quantification of timothy grass. The study was conducted using two controlled environment chambers (EGC Corp., Chagrin Falls, OH) at Beltsville, Maryland, with a given chamber set at one of four atmospheric [CO_2] concentration set points (300, 400, 500, and 600 [micro]mol/mol) for 24 hr/day. The atmospheric [CO_2] concentration of the air within each chamber was controlled by adding either [CO_2] or [CO_2]-free air to maintain the set concentration. Actual average 24-hr atmospheric [CO_2] concentration values ([+ or -] SD) were 315 [+ or -] 19, 395 [+ or -] 11, 491 [+ or -] 10, and 589 [+ or -] 12 [micro]mol/mol. Injection of [CO_2] and [CO_2]-free air was controlled by a TC-2 controller using input from an absolute infrared gas analyser. Chamber temperatures were altered diurnally (in steps) from an overnight low of 20°C to a maximum afternoon value of 30°C. Photosynthetically active radiation (PAR) was altered concurrently with temperature, with the highest PAR value (900-1,000 [micro]mol/[m^2] per sec) occurring during the afternoon (1,200-1,500). The duration of daily PAR was 14 hr, which was supplied by a mixture of high-pressure sodium and metal halide lamps.

Seeds of timothy grass (P. pratense, var. Climax) were obtained from Pawnee Buttes Seed (Greeley, CO). We sowed three seeds in each 4 inch pot used and thinned to one seedling per pot 6-8 days after emergence. Pots were watered to the drip point daily with a complete nutrient solution containing 14.5 mmol/L nitrogen. Ten pots were grown in each of the two chambers during each run. In eight runs over time, each level of [CO_2] concentration was run a total of four times, twice in each chamber. Plants from the third run were stunted and grew poorly for unknown reasons; this run was omitted, resulting in three runs for [CO_2] concentration levels of 300 and 500 and four runs for [CO_2] concentration levels of 400 and 600.

Plants were harvested 60 days after sowing. All leaves were separated from stems above the collar and weighed. Ten leaves from each plant were used for determination of total area, mass, and nitrogen and carbon content, which was determined using a PerkinElmer 2400 Series II CHNS/O analyser. Leaf subsamples of approximately 2 g fresh weight were removed from each plant for fungal growth, refrigerated, and processed within 3 hr of harvest.

All solutions, suspensions, and rinses were made using sterilised, deionised, distilled water (ddw) with Tween 80 added at the rate of 5 drops/L (ddw+T). Leaves were surface-sterilised by immersion and agitation in 1 L of 0.3% sodium hypochlorite solution for 3 min, followed by three successive 1-min rinses. All solutions and rinses were changed after leaves from all plants in a chamber were immersed to avoid transfer of any leaf material among treatments. Surface-sterilised leaves were oven-dried at 68°C for 48 hr.

Inoculum Preparation, Leaf Inoculation, and Fungal Growth

The fungal isolates grown were A. alternata (Fries) Keissler, isolate EGS 34-016, obtained from the collection of E. Simmons, and C. phlei (C.T. Greg.) G.A. de Vries, isolate CBS 306.5, obtained from the Fungal Biodiversity Centre of the Central bureau voor Schimmel cultures (Utrecht, the Netherlands). Isolates were cultured on standard petri plates with half-strength V8 juice agar (100 mL V8 juice; Campbell's Soup Company, Camden, NJ), 900 mL ddw, 2 g [$CaCO_3$], 15 g agar powder, autoclaved for 20 min at 121°C).

Fungal inocula were prepared immediately before inoculating leaves and prepared the same way for both fungal species. Two Petri plates with conidia covering most of the agar surface were selected. Conidia were removed from these plates by pouring small aliquots of ddw+T onto the surface of the agar in each plate, gently scraping the agar surface with a ceramic scraper, pouring the water with removed conidia through four layers of sterilised cheesecloth (Fisher Scientific) into a sterilised 1-L beaker, and repeating three more times. The resulting spore suspensions were vortexed at maximum speed for 30 sec to separate clusters of spores and subtending hyphae. The total volume was brought up to 1 L with additional ddw+T. Inoculum strength was standardised visually: A 10-mL subsample was removed with a glass wide-mouth pipette, filtered with a 47-mm diametre, 0.45-[micro]m pore size, gridded mixed-cellulose filter in a glass vacuum filtration apparatus, and examined at 5x magnification under a dissecting microscope. If spores were more than one layer deep on the filter, ddw+T was added to the incolumns suspension in 200 mL increments until subsamples resulted in filters densely covered with a single layer of spores. Five hundred millilitres of the final inoculum suspension was placed into two sterilised 1-L beakers; each beaker was used to inoculate leaves from a single chamber.

Weighed leaf subsamples (0.5 [+ or -] 0.075 g dry weight) from each single plant were inoculated separately by immersion and agitation in the inoculum suspension for 1 min. Inoculated leaves were placed in a sterilised media bottle (250-mL size used for A. alternata, 500-mL size for C. phlei) with caps placed loosely, then incubated for 1 week in an environmental controller set at 21°C with a 12-hr light/dark cycle and ambient [CO_2] concentration.

Spore harvest and A. alternata antigen extraction. After 1 week, 200 mL sterilised ddw+T was added to each media bottle. Bottles were vortexed at maximum speed for 30 sec to dislodge spores from leaves and subtending hyphae. For A. alternata only, antigen was then extracted: media bottles were left at room temperature for 2 hr; bottles were then inverted several times by hand. With a sterile syringe, 20 mL water with suspended spores was removed from each bottle and passed through a syringe filter (0.22-[micro]m pore size, 33-mm diametre, polyethersulfone; Millipore, Billerica, MA). The resulting spore-free extract was frozen inside a 50 mL plastic centrifuge tube and lyophilised on a Freezone 4.5 freeze dry system. For both fungal species, media bottles were then inverted several times by hand; 10 mL sub-samples were removed using glass wide-mouth pipettes, placed on filters, and examined as described above. If spore density on the filter was too sparse (or too crowded) for accurate counting, a larger (or smaller) subsample was obtained, a new filter was made, and spore density was rechecked. The final volume used for each subsample ranged from 1 to 100 mL and was recorded for extrapolation of the total number of spores produced per gram of dried leaf tissue.

Spore quantification. Filters were rewetted from below with several drops of sterilised ddw and placed on a compound light microscope stage. Eight 300 x 400 [micro]m fields of view were digitally photographed (Axiocam digital camera and Axiovision imaging software; Carl Zeiss Microimaging, Inc.) at random positions over one half of each wetted filter. Spores within each of the eight fields were counted manually from the photographs; all whole, uncollapsed spores were counted regardless of size. For A. alternata only, the first 10 spores laying flat on the filter were also measured for length and width.

Antigen Quantification

Lyophilised extracts were reconstituted in 3 mL ddw and assayed for A. alternata antigen using a competitive inhibition ELISA described by Salo et al. (2005). The assay kit consisted of a rabbit polyclonal antibody (lot ZA4-4; Greer Laboratories, Lenoir, NC) produced using whole mycelial extracts of A. alternata (lot XPM1-X10); the A. alternata mycelial extract was also used as the standard in the assay. This assay has been shown to detect multiple A. alternata antigens including the allergen Alt a 1. Briefly, the standard A. alternata extract was bound to plastic microtiter plate wells, excess was washed away, and unbound sites were blocked with 1% bovine serum albumin solution. Dilution series of the sample extracts were pipetted into the wells and followed immediately by the rabbit anti-A. alternata antibody. After incubation, we aspirated solutions, washed the wells, and added a per-oxidase-rabelled goat anti-rabbit antibody solution (Sigma-Aldrich, St. Louis, MO). Finally, we added substrate and measured the colour change at 405 nm using a microtiter plate reader (Molecular Devices, Sunnyvale, CA). We compared the resulting reaction rates against a dilution series of standard A. alternata antigen to determine antigen concentration. Results are reported as micrograms A. alternata antigen per spore or per plant.

All statistical analyses were conducted using R version 2.8.1. The R package "lme4" (Bates et al. 2008) was used to fit and evaluate generalised linear mixed models of the following effects: atmospheric [CO_2] concentration level (fixed at 300, 400, 500, or 600 [micro]mol/mol), run (random effect of seven runs over time), and chamber (random effect of the two growth chambers used). All fungal responses were also modelled with the covariate of leaf C:N of each plant. Factors were sequentially removed from models and model fit was evaluated using Akaike's information criterion; for each response, the simplest statistical model with good fit was used. We used the R package "language R" to estimate p-values for the fixed effect of atmospheric [CO_2] concentration level in each model using Markov

Chain Monte Carlo resampling and to generate highest posterior density confidence intervals (CIs) for means at each atmospheric [CO_2] concentration level. We considered p-values [less than or equal to] 0.05 to be statistically significant. The responses evaluated were the following: a) A. alternata spore counts (natural log transformed to meet statistical model assumptions that variance does not increase with group mean); b) mean length and width of A. alternata spores; c) quantity of antigenic protein produced (square-root transformed to linearise relationships with model factors and to meet statistical model variance assumptions); and d) C. phlei spore counts (natural log transformed, as above).

The random effect of chamber was neither significant nor needed to improve model fit for any responses and is not discussed further. The random effect of run was very important for modelling all plant and fungal responses; model-adjusted means, which account for variation due to run, are presented for responses at individual levels of [CO_2] concentration. When the fixed effect of [CO_2] concentration was determined to be statistically significant, the R package "multcomp" was used to run pairwise Tukey's comparisons of [CO_2] concentration level means. The contribution of each factor toward explaining response variance was calculated as described by Kramer (2005).

Supplemental Experiment with Field-collected Plant Material

Nine samples of plant material were collected during routine mowing from three different sites at each of three local golf courses. A mixture of perennial ryegrass (Lolium perenne L.) and annual poa (Poa annua), both cool season grasses with [C_3] photosynthetic metabolism, is grown at all sites, but fertilizer application rates vary. The samples were collected on 8 January 2008 and refrigerated within 3 hr of mowing. Leaves were treated and used as growth substrate for the two fungal species as described above. Resulting C:N and spore counts were not statistically analysed because of small sample size but are shown for visual comparison.

Plant Responses to Atmospheric [CO_2] Concentration Levels

Leaf C:N was significantly higher in plants grown at 500 and 600 [micro]mol/mol atmospheric [CO_2] concentrations than at the two lower concentrations ($p = 0.017$). Leaf dry weight per plant was significantly greater at 600 [micro]mol/mol atmospheric [CO_2] concentration than at the three lower concentrations ($p < 0.001$). Specific leaf area was not significantly affected by atmospheric [CO_2]

concentration (nor shown). Plant responses were variable among individual plants and runs. Among individual plants from all treatment levels, leaf C:N ranged from a minimum of 7.7 to a maximum of 16.2, which corresponds to total leaf nitrogen contents of 5.6% and 2.8%.

Fungal sporulation on leaves grown at four atmospheric [CO_2] concentration levels. The log-transformed counts of A. alternata spores produced per gram leaf were positively correlated with leaf C:N ($p < 0.001$; adjusted [R^2] = 0.25), with an order-of-magnitude increase across the range of leaf C:N of experimental plants. A. alternata sporulation on field-collected grass leaves (supplemental experiment) shows a similar relationship. Mean length and width of A. alternata spores did not change significantly with leaf C:N. In contrast to the positive association with spore numbers, the quantity of A. alternata antigen per spore (square-root transformed) decreased as leaf C:N increased ($p < 0.001$; adjusted [R^2] = 0.34). Despite the decreased antigen content per spore at higher C:N, the quantity of A. alternata spores as well as the total antigen produced on a per-plant basis were positively associated with atmospheric [CO_2] concentration levels beyond the variation attributed to leaf C:N, with nearly three times the number of spores and twice the amount of antigen produced on plants grown at 500 and 600 [micro]mol/mol atmospheric [CO_2] concentrations ($p < 0.001$) than at the lower two concentrations. The log-transformed counts of C. phlei spores produced per gram leaf also increased with C:N ($p < 0.001$; adjusted [R^2] = 0.22). Spores of C. phlei were produced in much lower numbers than A. alternata spores, although the slopes of the relationships with leaf C:N were not significantly different between the two fungi ($T = 1.21$; $p > 0.20$). In contrast with A. alternata spores, however, the mean number of C. phlei spores produced on a per-plant basis was not significantly different among plants grown at different [CO_2] concentration levels.

Climate change and urbanisation are expected to increase the prevalence of asthma and allergies. Concomitant rises in temperature, atmospheric [CO_2] concentration, and pollen abundances over recent decades have been suggested as causes of increasingly prevalent and severe asthma and allergy symptoms observed over the same time period. Our findings indicate that, as with pollen production, the sporulation of allergenic fungi is likely to be amplified as atmospheric [CO_2] concentration increases and therefore is also likely to contribute to increasing prevalence and severity of asthma and allergies. We found significant positive relationships between leaf C:N and A.

alternata and C. phlei spore production per gram leaf, although only A. alternata sporulation was associated with [CO_2] concentration level after accounting for changes in leaf C:N. These relationships suggest that considerable increases in sporulation of both species will occur, and may already be occurring, if rising atmospheric [CO_2] concentration increases plant C:N in the field. The response of C. phlei, however, may not be as strong as that of A. alternata. More study of this species is needed.

Although C:N was correlated with increased sporulation, a large proportion of the variability in spore production remains unexplained. Some of this variability in sporulation may result from inherent variation in growth among individual plants, which can be of a magnitude large enough to obscure responses to elevated atmospheric [CO_2] concentration treatments. Other aspects of environmental variability, which deserve additional scrutiny, may also contribute to variation in spore numbers and quality. Temperature was kept constant during fungal growth in this experiment, but relative humidity was not controlled. Additional plant attributes such as leaf tissue concentrations of mineral elements (other than nitrogen), stomatal density, and other leaf surface properties, and the chemistry of leaf carbon compounds were not measured. All of these factors may contribute to the variation in sporulation, and all are also likely to be altered by anthropogenic climate change. For example, plant transpiration is likely to decrease under elevated atmospheric [CO_2] concentration, which often leads to increased moisture in soils and overlying litter; warming temperatures and altered precipitation patterns will further impact moisture in soil and plant litter layers. Global climate change factors encompass many direct and indirect effects on plants and fungi, as well as complex interactions and feedbacks. The net effects of global changes are difficult to capture with a single study. Long-term (multidecadal) observational studies are needed to distinguish the impact of multiple abiotic changes on allergenic spore production in toto.

Our results corroborate the findings of Klironomos et al. (1997), who found large increases (~ 100-150%) in Alternaria spore abundance in chambers with poplar trees growing in elevated atmospheric [CO_2] concentration. Our results also suggest the potential ubiquity, across plant species, of increased spore production as a function of increasing atmospheric [CO_2] concentration. Klironomos et al. (1997) also found a large increase in the abundance of spores from the genus

Cladosporium at elevated atmospheric [CO_2] concentration (~ 225-350% increases), which was not apparent in our study. A strong response to atmospheric [CO_2] concentration may relate to a saprobic (growing on dead plant tissues) lifestyle. A. alternata is facultatively pathogenic or saprobic. Although C. phlei can grow on dead plant tissues, as it did in this experiment, it is described primarily as a pathogen of living timothy grass. Other species within Cladosporium, however, are saprobic. Specialisations that allow fungal species to grow on specific host plants, such as unique metabolic pathways, may cause these fungi to be limited more by other plant characteristics than by carbon content alone.

Although only a few studies have examined fungal sporulation on plants grown at elevated atmospheric [CO_2] concentration, there has been extensive work examining sporulation on artificial growth media and on plants with different levels of nitrogen content. In some pathogenic fungal species, sporulation on living leaves is greatest at medium C:N, but other species sporulate maximally at high C:N. The C:N that promotes the most sporulation varies among fungal species and even strains. These varying trends among fungal taxa may reflect different reproductive strategies, specialisations, and functional tradeoffs needed to thrive in different environmental and host conditions. There is experimental evidence that while the greatest numbers of spores may be produced at high C:N in some fungal taxa, spore protein content, germination success, and host-plant infectivity are greater when more nitrogen is available. In our study, the decreasing content of antigenic protein in A. alternata spores grown at higher C:N suggests increasing nitrogen limitation of the fungus. It is not yet clear if this nitrogen limitation is associated with compromised germination, growth, or aller-genicity of individual A. alternata spores.

Meta-analyses of plant responses to elevated atmospheric [CO_2] concentration reveal clear trends despite variation among plant species and experimental conditions. The mean increase in C:N across many studies of grass species grown at doubled atmospheric [CO_2] concentration was 24.4%. The responses of plants to less-than-doubled increases in atmospheric [CO_2] concentration have not been well studied, so expectations for C:N of plant tissues grown in real-world, gradually increasing atmospheric [CO_2] concentration cannot be predicted with certainty. However, an observational study of herbarium leaf specimens found that leaf nitrogen content decreased by a mean of 17% across samples from the early, mid, and late 20th century for 11 [C_3] plant species [recorded global levels of atmospheric [CO_2] concentration increased from < 320 in 1958 to 388 [micro]mol/mol in July 2009 (National Oceanic and Atmospheric Administration 2009)]. The authors attribute the decreased leaf nitrogen content to the combination of direct and indirect (warming) effects of rising atmospheric [CO_2] concentration, and their findings are corroborated by other observational and experimental measurements. Thus, this evidence of increased plant C:N associated with rising atmospheric [CO_2] concentration and climatic change is consistent with increased sporulation of allergenic fungi, as observed in the current study, and has implications for increasing the prevalence and severity of allergy and asthma symptoms.

Ultrafine Manganese Oxide Particles

An important step in assessing the toxicology of particles is to determine their fate after inhalation. Of particular interest to us are airborne ultrafine particles (UFPs; < 100 nm), which are abundant in ambient urban air and are of the same size as engineered nanoparticles. Translocation to extrapulmonary sites after respiratory tract deposition represents an important mechanism for these particles to cause direct effects in secondary target organs. The extent to which this process occurs depends on several factors including particle solubility, particle or aggregate size, the site of deposition, and the integrity of the epithelial lining. UFPs deposit efficiently in all regions of the respiratory tract, depending on their size; specifically, as particle size decreases toward the smallest UFPs, nasopharyngeal deposition increases.

Studies in rats have shown translocation of soluble manganese compounds from the nose along olfactory neuronal pathways to the olfactory bulb after inhalation or intranasal instillation exposures. Likewise, the few studies that have examined the fate of UFPs deposited on the nasal mucosa identified translocation along the neuronal olfactory route as a pathway to the olfactory bulb of the central nervous system (CNS). These include early studies in nonhuman primates, which demonstrated the translocation of solid nanosized particles (30 nm poliovirus; 50 nm silver-coated gold colloids) along the axons of olfactory nerves into the olfactory bulb. We have also shown that inhaled elemental carbon particles ([^{13}C]; 35 nm, count median diametre) accumulate in rat olfactory bulb after whole-body inhalation. Regarding penetration into deeper brain regions, Tjalve et al. (1995) demonstrated that soluble ionic Mn instilled into the olfactory chamber of pike has the ability to pass synaptic junctions and migrate from the olfactory tract to more distal regions, including the hypothalamus. Dorman et al. (2004) found Mn in the striatum and cerebellum of rats after subchronic inhalation exposure to a soluble Mn salt (sulphate); however, this was attributed to uptake from the blood. Thus, contributions to brain Mn levels from the blood need to be considered and may also be an issue for inhaled solid UFPs.

The effects of translocated particles in the brain are also important to determine. For example, preliminary information has emerged from populations of welders that some of them may develop parkinsonism 17 years earlier than the general population. Welding produces high amounts of fumes containing Mn UFPs. Several recent epidemiologic studies describe occupational exposure ranges of approximately 0.01-5 mg/$[m^3]$ Mn in fumes from various welding processes and materials. Conflicting data emerge from animal studies, however, regarding effects of inhaled Mn compounds in the brain. Henriksson and Tjalve (2000) reported changes in glial fibrillary acidic protein (GFAP) and S-100b, markers of astrocyte activation, in several brain regions from rats exposed intranasally to Mn chloride. However, Dorman et al. (2004) did not find any evidence of changes in GFAP levels in the brain after exposure to Mn sulfate or phosphate. Potential contributing factors to the lack of concurrence in results include differences in the solubilities of the Mn salts used, the doses, and the contribution of olfactory epithelial damage.

In the present study, we sought to address the hypothesis that a major translocation route for inhaled poorly soluble Mn oxide UFPs from deposits in the nose is to the olfactory bulb in the CNS. We characterised the size, oxidation state, and in vitro solubility of gas-phase-generated Mn oxide particles and also compared the translocation kinetics to the olfactory bulb of Mn oxide and $Mn[Cl_2]$ that were applied to the nasal epithelium of rats via instillation. We then measured the accumulation of Mn in lung, liver, and olfactory bulb after repeated inhalation exposures with both nares patent or with one naris occluded. We show that Mn oxide UFPs are translocated to and retained in the olfactory bulb (ipsilateral to the patent naris only) and present evidence of exposure-induced effects in that region of the brain. These studies demonstrate the importance of UFPs size and of solubility in olfactory translocation processes.

Specific pathogen-free male Fischer 344 rats (200-250 g body weight, 3 months of age) were obtained from Harlan (Indianapolis, IN) and housed in filter-top plastic cages in a facility accredited by the International Association for Assessment and Accreditation of Laboratory Animal Care; animals had free access to Purina rodent chow and water. The background concentration of particles in this facility is extremely low (< 50 particles/[cm^3]). All animals were allowed to acclimate for at least 1 week before use in experimental protocols, which were approved by the University Committee on Animal Resources (University of Rochester). Animals were treated humanely and with regard to the alleviation of suffering.

Generation of Mn Oxide UFPs

We generated Mn oxide UFPs (~500 [micro]g/[m^3]; 18 x [10^6] particles/[cm^3]) via electric arc discharge in an argon-filled chamber between two opposing Mn rods (purity, 99.95%; Electronic Space Products International, Ashland, OR). Oxygen was introduced into the generator (20 mL/min) to ensure the formation of the metal oxide. Particle electrostatic charge was brought to Boltzman equilibrium ([^{210}Po] source). Available instrumentation monitored particle mass and number concentrations continuously as well as aerosol size distribution at regular intervals [tapered element oscillating microbalance; condensation particle counter, model 3022A, and scanning mobility particle sizer, model 3080].

Gas-phase-generated Mn oxide UFPs were collected for electron microscopic analyses immediately after generation using a point-to-plane electrostatic precipitator for transmission electron microscopy (TEM) samples (In-Tox Products, Moriarty, NM). For scanning electron microscopy (SEM) analysis, the particles were suspended in methanol, one drop placed on an SEM stub and desiccated for 24 hr. Electron micrographs showed that the aerosol was composed of agglomerates with an equivalent sphere diametre of approximately 30 nm. Agreement between scanning mobility particle sizer-derived and TEM-derived size distributions was very good. Primary particles varied between 3 and 8 nm in diametre, depending on the generation conditions.

Exposure of Rats to Mn Oxide UFPs

For whole-body inhalation exposures, rats were placed in compartmentalised, horizontal flow chambers (31-L Lucite tank; airflow = 30 L/min). In some exposures, the right naris was occluded according to methods outlined by Brenneman et al. (2000). Briefly, a 3 mm piece of polyethylene tubing [0.024 in. outer diametre (OD)] was inserted into an 8 mm piece of Silastic tubing (0.065 in. OD), the ends of which were sealed to form smooth, round edges. One day before exposure, rats were lightly anesthetized with Halothane, and the plug was inserted into the right naris; a small amount of Duro Quick Gel ensured that the plug remained in place. Inhalation exposures were for 6 hr/day, 5 days/week for up to 12 days with both nares open (12 days total for tissue Mn determinations; 11 days total for gene and protein array analyses). With one naris occluded, the exposure was for 2 days.

For intranasal instillation exposures, the particles were collected on a filter and resuspended in sterile pyrogen-free saline with sonication. Because nasally instilled material is readily aspirated into the lower respiratory tract in rodents, thus making olfactory mucosa exposure less than optimal, we developed a simple method to maximise the dose to the olfactory mucosa. Rats were lightly anesthetized with Halothane (3%).

The trachea was visualised with a pediatric otoscope and cannulated transorally with the plastic sheath of a 14 gauge catheter. The plane of anesthesia was maintained with Halothane (1.5%). Breathing through the tracheal cannula ensured that aspiration of the nasal instillate into the lung did not occur. A 30-gauge needle covered with polyethylene tubing was attached to a 1-mL syringe and the tubing was inserted into the left naris (5 mm), after which 5-7 [micro]g (in 30 [micro]L) of the suspended particles was slowly injected with the rats in a supine position. The rats were kept supine for 5 min, after which the Halothane was turned off and the tracheal cannula was removed once the animal regained consciousness. These methods were used for both Mn oxide and Mn[Cl_2] exposures.

Characterisation of the Oxidation State and Oxide form of Mn Oxide UFPs

We investigated the oxidation state of sampled Mn oxide particles using electron energy loss spectroscopy and an analytical TEM/STEM operating in scanning transmission electron microscopy (STEM) mode. Mn L-electron and oxygen K-electron energy loss edges were recorded from discrete primary particles and clusters of particles. We estimated elemental stoichiometry through comparison of the integrated area under energy loss edges after background subtraction, using Digital Micrograph (Gatan Inc.). Mn oxidation state was further characterised by comparing acquired spectra with published Mn L-electron X-ray absorption spectra for Mn oxide (MnO), Mn tetroxide ($[Mn_3][O_4]$), manganese sesquioxide ($[Mn_2][O_3]$), and Mn dioxide ($Mn[O_2]$). Further information on the calculation of the relative percentages of each state is given in Supplemental Material available online.

Dissolution Rate of Mn Oxide UFPs

We employed two methods to determine the dissolution rate of the gas-phase-generated Mn oxide UFPs in solution: ultrafiltration and dialysis. For ultrafiltration, samples (0.5 mg) were dispersed in Mn-free physiologic saline (1 mL, pH 7.4) using an ultrasonic bath. Sample suspensions were injected into the inlet flow line of the ultrafiltration sample cell fitted with a 1,000 molecular weight cutoff membrane. The sample was washed into the cell with an additional 2 mL saline solution. For dialysis, the sample suspensions were injected into the upper chamber of a dialysis cell fitted with a 1,000 molecular weight cutoff membrane. The sample cells were then swirled to evenly disperse the suspensions; after several hours, the suspensions settled as an even layer on the membrane filters. Inlet ports were connected to a peristaltic pump for optimal flow; the outlet ports were connected to a waste bottle. Outlet lines were periodically connected to sample tubes, and 50 mL of the filtrate or dialysate was collected and analysed for Mn by inductively coupled plasma optical emission spectroscopy (ICP-OES). For measurements of the dissolution rate above room temperature, the entire sample cell was immersed in a water bath at the desired temperature.

Cellular and Biochemical Parameters in Bronchoalveolar Lavage Fluid

Animals were euthanised with an overdose of sodium pentobarbital (intraperitoneal, 50 mg/100 g body weight) followed by exsanguination. As described in detail elsewhere, the lungs, trachea, and heart were removed en bloc, and the right lungs were lavaged after weighing with a fixed volume of sterile, pyrogen-free 0.9% saline (five times, each with 5 mL), separating the first two lavages for protein and enzymatic analyses. Cells were pelleted by centrifugation (400 x g) for 10 min. The cells were pooled from all lavage fractions for viability determination (trypan blue exclusion), enumeration, and differential analysis. Total protein concentration and lactate dehydrogenase and [β]-glucuronidase activities were measured using commercially available kits.

Analyses of Metal Content in Tissue Samples

At the time of sacrifice, we removed the skin and fur from the rats with a dedicated set of instruments, and the carcass was thoroughly washed before the removal of tissues; the carcasses were then moved to a separate room for excision of organs to be analysed for metals content with different sets of instruments. The left lungs, 1 g of liver tissue, and several brain regions (olfactory bulbs, striatum, trigeminal ganglions, midbrain, frontal cortex, cortex, cerebellum) were removed to measure their Mn and iron content. In the intranasal instillation studies, right and left olfactory bulbs and mucosa (including turbinates and cribriform plate) were removed. We did not perfuse the tissues because preliminary tests showed that perfusion did not affect Mn levels. The tissues were placed directly into Teflon digestion vials, weighed, and wet ashed with ultrapure 70% nitric acid. After ashing, the tissue residue was resuspended in HN[O_3] and the concentration was adjusted to 2% with 18 M[ohm] deionised water before graphite furnace atomic absorption spectroscopy analysis. Quantitation was achieved through comparison to reference standards.

RNA preparation for array and blot analysis. Tissue samples were prepared as previously described using the standard RNAzol protocol for array and blot analysis. RNA was amplified using the primer-specific Smart Probe Amplification Kit designed specifically for the Atlas array systems. Briefly, 50 ng of total RNA was amplified using specific primer sets for reverse-transcriptase polymerase chain reaction (PCR) amplification to full length double-stranded cDNA, which ensures amplification of representative original gene population. Double-stranded cDNA generated from the PCR amplification step was pooled separately from control and treated groups, respectively, and cDNAs were prepared with standard reagents and the Rat 1.2 and Rat 1.2 II Array kits (Clontech) following the manufacturer's instructions. Rabelled cDNA was generated during the first strand synthesis supplemented with [[α]-[^{32}P]]-deoxyadenosine 5'-triphosphate (3,000 Ci/mmol, 10 [micro]Ci/[micro]L) using Moloney Murine Leukemia Virus (MMLV) reverse transcriptase (Clontech) and purified by column chromatography.

Protein Array Aanalysis

Protein was extracted from lung and brain samples using the BD Clontech Protein Extraction and Rabelling Kit (Clontech). Briefly, 1 [micro]g of protein from each sample was pooled and used to probe the RayBio Rat Cytokine Array I according to the manufacturer's protocol. After 20 min exposure, the membranes were scanned using a phosphorimager and analysed for relative intensity of expression. Data are expressed as fold change differences from the unexposed controls.

Statistical Aanalyses

We analysed results for statistical differences by one-way analysis of variance with appropriate data transforms using SigmaStat (Systat Software Inc., Point Richmond, CA). Data were appropriately transformed if an analysis of residuals suggested deviations from the assumptions of normality and equal variance. Differences between groups were further analysed using Tukey multiple comparisons. Such comparisons were considered statistically significant when p [less than or equal to] 0.05.

Tissue distribution of Mn and effects in the lungs and brain after repeated inhalation exposure to ultrafine Mn oxide aerosols. We exposed rats to either filtered air or to ultrafine Mn oxide aerosols (465 [+ or -] 94 [micro]g/[m^3]) for 6 hr/day for 12 days. There were no indications in terms of either cellular or biochemical parameters to indicate that active lung inflammation occurred as a result of exposure; the only statistically significant changes found were decreases in lavage fluid total cell number after 6 days of exposure and [β]-glucuronidase activity after 12 days of exposure. The Mn content in lung tissue increased progressively and significantly after 6 and 12 days of ultrafine Mn oxide aerosol exposure, representing about a doubling of the tissue Mn content. Inhalation of Mn oxide aerosols for 12 days also resulted in significant increases (~3.5-fold) in olfactory bulb Mn content. Significant increases in Mn content were also found in the striatum, frontal cortex, and cerebellum after 6 and 12 days of exposure and in the cortex after 12 days of exposure, although the magnitude of these changes was much lower than what was observed in olfactory bulb. The increase in liver Mn content was smaller than the increase in lung tissue (controls, 2.78 [+ or -] 0.04 ng Mn/mg tissue; 6 days, 2.61 [+ or -] 1.61 ng Mn/mg tissue; 12 days, 2.91 [+ or -] 0.09 ng Mn/mg tissue). The tissue concentration of Fe also increased slightly, but significantly, after exposure in lung, olfactory bulb, and cerebellum.

Gene microarray analyses were performed on tissue samples obtained from various brain regions after a total of 11 days of ultrafine Mn oxide aerosol exposure (498 [+ or -] 69 [micro]g/[m^3]). Protein expression of TNF-[α] correlated with gene expression increases. Several other genes involved in inflammation (e.g., macrophage inflammatory protein-2) and stress responses (e.g., GFAP) also showed 2-fold or greater increases in expression over controls in the olfactory bulb. Although most of the changes occurred in olfactory bulb tissue, neuronal cell adhesion molecule was also increased in the frontal cortex and midbrain. Several other genes exhibited increases in expression that were slightly less than 2-fold higher compared with controls.

Tissue distribution of Mn after inhalation of ultrafine Mn oxide aerosols with one naris occluded. The data presented above show that Mn accumulation was the highest in the olfactory bulb and lower in other brain regions, consistent with olfactory translocation after repeated inhalation exposures; however, the Mn in those tissues could have arisen from bloodborne Mn. In a separate set of experiments, rats were exposed with the right nares occluded for 6 hr on each of 2 consecutive days. Three animals each were killed for tissue Mn analyses immediately after the first exposure and 18 hr after the first and second exposures. Control animals with both nares open were killed 18 hr after a 6 hr exposure; there were also unexposed controls (0 hr). With both nares patent, significant increases in lung Mn content were found 18 hr after a 6 hr ultrafine Mn oxide aerosol exposure, as was expected from the repeated inhalation studies; liver Mn content also increased. When the right naris was occluded during exposure, lung tissue Mn content increased 11 fold to 1.9 [+ or -] 0.05 ng/mg wet tissue immediately after the 6 hr exposure; liver Mn content increased from 2.5 [+ or -] 0.01 to 2.9 [+ or -] 0.08 ng/mg tissue. After 18 hr of recovery, significant amounts of Mn remained in the lung (21% of the amount deposited at the end of exposure), but most of the deposited material had been cleared, and no increase was observed in liver tissue. Two consecutive days of exposure (6 hr each) did not significantly alter the retention kinetics of the ultrafine Mn oxide aerosols deposited in the lung. Thus, clearance of Mn from the lung under these exposure conditions is rapid.

Mn also accumulated in the left and right olfactory bulbs when both nares were patent. However, when the right naris was occluded, Mn accumulated only in the olfactory bulb on the patent (left) side. Furthermore, Mn accumulation increased in left olfactory bulb tissue, unlike the lung, with time postexposure and exposure duration. A small, but insignificant, amount of Mn also accumulated in the right olfactory bulb after two consecutive 6 hr exposures (1.2-fold increase), potentially from bloodborne Mn due to dissolution of the Mn oxide in alveolar macrophages or due to transport between the nares via a small perforation in the rat nasal septum, as has been previously described.

Mn Oxide UFP Oxidation State and Solubility

EELS analysis of primary particles and agglomerates indicated a mean stoichiometry close to MnO. Comparison of Mn L-electron edge structure with published spectra for different oxidation states indicated that the particles were composed of MnO [61%, Mn(II)] and [Mn_2][O_3] [39%, Mn(III)]. At room temperature, both the ultrafiltration and dialysis experiments showed that the Mn oxide UFPs dissolved at a rate of 1-1.5% per day (i.e., Mn detected in outflow via ICP-OES) at a neutral pH similar to the nasal mucosal milieu. Subsequent dialysis experiments also showed that temperature, solution flow rate, and time (up to 10 days) did not affect the dissolution rate. However, acidification to pH 4.5, similar to the phagolysosomal conditions of alveolar macrophages, resulted in rapid dissolution. This soluble fraction most likely contributed to bloodborne Mn that was then distributed throughout the body.

Olfactory bulb uptake of Mn after intranasal instillation of solid ultrafine Mn oxide or a soluble Mn salt. The issue of whether or not the solubilisation of ultrafine Mn oxide was a prerequisite for its translocation along the olfactory nerve was still outstanding after the experiments described above. Given the low solubilisation rate (1-1.5% per day) of ultrafine Mn oxide at neutral pH, one would predict that only this small fraction of Mn oxide deposited on the olfactory mucosa would be translocated to the olfactory bulbs in its soluble form. In order to test this, we instilled 30 [micro]L of soluble Mn[Cl_2] (5.3 [micro]g Mn) or saline-suspended Mn oxide UFPs (6.7 [micro]g Mn) into the left naris of anaesthetized rats. Despite the fact that only 1.5% of the Mn in the oxide form was soluble at a maximum, similar Mn burdens (given as a percentage of the instilled Mn) were found in the left olfactory bulb tissue 24 hr after intranasal instillation of the chloride (8.2 [+ or -] 3.6%) or the oxide (8.2 [+ or -] 0.7%). Furthermore, less Mn from the oxide was found on the cribriform plate area and the turbinates (representative of the olfactory mucosa), indicating greater retention of the Mn[Cl_2] by those tissues. In another group of rats, olfactory bulb uptake of intranasally instilled Mn oxide occurred rapidly (within 30 min) after exposure, although to a very small degree (0.2% of amount found on olfactory mucosa at 30 min); however, this had increased to 6.8% after 24 hr. In another experiment, Mn oxide UFPs that were instilled into the left naris translocated primarily to the left olfactory bulb, although a small amount also appeared in the right olfactory bulb, possibly due to some exposure of the right side of the nose via the nasal septal window. These data indicate that the appearance of Mn in olfactory tissues cannot be due to soluble Mn but, rather, that solid UFP transport occurs efficiently.

Traditionally, the respiratory tract is considered a target organ for effects of inhaled solid particles. However, more recent evidence from epidemiologic, controlled clinical and animal studies with ambient particulate air pollutants shows that extrapulmonary organs are also affected. Specifically, it has been hypothesized that inhaled UFPs accumulate and cause effects in extrapulmonary organs, such as the cardiovascular system and CNS because of their propensity to translocate across epithelial barriers. Indeed, in our present study we demonstrate that 6- to 12 day inhalation exposure of rats for to solid Mn oxide UFPs resulted in significant increases of Mn in several brain regions, most notably the olfactory bulb. The fact that occlusion of the right naris during inhalation for 2 days of the nanosized Mn oxide led to accumulation of Mn only in the left olfactory bulb confirmed that translocation from nasal deposits along the olfactory nerve accounts for this increase.

The olfactory translocation route has been well demonstrated for soluble Mn compounds, and it has been suggested that solubility is an important determinant of the efficiency of this process. Dorman et al. (2004) measured the translocation of poorly soluble Mn phosphate into olfactory bulb, cerebellum, and striatum. In that study, only the olfactory bulb, and not the more distal structures, demonstrated a small increase in Mn content. One important difference to our study, however, is that the phosphate particles had a mass median aerodynamic diametre of 1.6 [micro]m, whereas the count median diametre of the Mn oxide particles used here was 31 nm. The axons of olfactory neurons narrow to a diametre of approximately 200 nm and become tightly packed where they pass through the cribriform plate pores. Thus, in order for solid particles deposited on the olfactory mucosa to be transported through the cribriform plate, they should be < 200 nm in size. We suggest, therefore, that particle size plays a most important role and that dissolution is not a prerequisite for neuronal uptake and translocation of solid UFPs in the absence of mucosal injury. Indeed, this suggestion is based on earlier studies in nonhuman primates by Bodian and Howe (1941a, 1941b) and by DeLorenzo (1970) with intranasally instilled 30 nm poliovirus and 50 nm silver-coated gold particles, respectively, and on our recent study in rats with inhaled carbon-13 particles (median size, ~30 nm) showing translocation along olfactory neuronal pathways.

The Mn oxide UFPs used in the present study dissolved at a rate of about 1.5% over 24 hr in physiologic saline at neutral pH. However, greater solubilisation of Mn oxide has been demonstrated at approximately pH 4.5, as would be encountered in the phagolysosome of alveolar macrophages. Given that the nasal mucosa has a pH close to neutral and free macrophages are not normally present on the nasal epithelial surface, rapid solubilisation of the intranasally instilled Mn oxide UFPs in our study is unlikely. We conclude, therefore, that the increase in olfactory bulb Mn in our study is due to neuronal uptake and translocation of the solid Mn oxide UFPs. Support for this suggestion comes also from the results of our comparison of intranasally instilled Mn oxide to soluble $Mn[Cl_2]$. If solubilisation of the Mn oxide UFPs were a prerequisite for neuronal translocation of Mn, one would expect that significantly less Mn would accumulate in the olfactory bulb for instilled Mn oxide compared with $Mn[Cl_2]$ because of the low solubilisation rate of the oxide. Instead, it was the same for both compounds. In addition, one would expect more of the Mn oxide than the chloride to be retained on the olfactory mucosa; however, the opposite was found. Finally, the accumulation kinetics are consistent with a rapid translocation velocity of solid particles in axons.

Another issue to consider is that Mn oxide UFPs that deposited in the alveolar compartment of the rat respiratory tract in our study may have undergone dissolution in alveolar macrophages followed by diffusion of soluble ionic Mn into the blood circulation. In addition, Mn oxide UFPs may translocate across the alveolar-capillary barrier and the blood-brain barrier in regions where it is discontinuous (e.g., choroid plexus, ventricles, brain stem, hypothalamus). Dissolution and absorption of Mn oxide particles deposited in the respiratory tract and cleared into the GI tract via mucociliary action could also contribute to bloodborne Mn; however, this will only be a small fraction, < 5% of the ingested amount. Although it is likely that bloodborne ionic Mn contributed to the increase in cerebellar Mn, the degree to which this may have contributed to Mn in other brain regions is not known. It is also possible that Mn crosses the blood-brain barrier via competition with Fe; both Fe and Mn can complex with transferrin in the blood and thereby compete for uptake into the CNS capillary endothelium via the transferrin receptor. In addition, Mn appears to facilitate the transport of Fe into the brain after chronic exposure. Our finding of higher Fe levels in olfactory bulb and cerebellum after Mn oxide inhalation may be related to competitive or facilitated transport, but this needs to be investigated in further studies (Fe was not present in the exposure atmosphere). Because the tissues were not per-fused, some of the Fe could have come from the blood, and future studies should be done using perfused tissues.

Frontal cortex and striatal Mn increases in our study may have been due either to a contribution from bloodborne Mn or to neuronal translocation from olfactory bulb to more distal brain structures. Large molecules, approaching the size of nanoparticles (wheat germ agglutinin-horseradish peroxidase complex), were found to translocate from the nasal cavity to the olfactory bulb and more distal structures by crossing synapses, as was also observed with herpes virus. Thus, although ionic bloodborne Mn can contribute to increased Mn levels in some brain regions, translocation of solid Mn oxide UFPs from nasal deposits to CNS regions via the olfactory nerve should not be excluded as a possible mechanism.

Using a predictive particle deposition model for the rat, we estimated from our data that about 11.5% of the amount deposited on the olfactory mucosa (369 ng) was translocated to the olfactory bulb (42.5 ng). Model inputs included the Mn oxide particle size distribution and exposure concentration; the default breathing parameters for rats (tidal volume, 2.1 mL; respiratory rate, 102/min) and the rat-specific nasal parameters were also considered. Olfactory bulb weights and Mn concentrations were those from experimental results; the olfactory bulb Mn background level in unexposed rats was subtracted from the level in rats exposed to Mn oxide for 6 hr with both nares patent.

The results of this study raise a number of important questions, including extrapolation to humans, the significance for neurotoxicity, implications for other inhaled nanosized particles, and translocation from olfactory bulb to deeper brain structures. Human exposures to high concentrations of Mn oxide-containing UFPs occur under certain occupational settings, such as arc welding. Concentrations of UFPs containing Mn oxide can reach [10^7] particles/[cm^3]. The issue of the neurotoxicity of inhaled Mn-containing particles is important insofar as welding fumes contain 0.01-5 mg/[m^3] Mn, depending on the welding process and materials used, and preliminary results from a small cohort of welders show that they may develop Parkinson-like symptoms. Dorman et al. (2004) showed that the inhalation of soluble Mn sulfate aerosols did not lead to an increase in GFAP protein expression in olfactory bulb tissue; however, Henriksson and Tjalve (2000) performed three successive weekly intranasal instillations of $Mn[Cl_2]$ and showed an increase in this protein, which is a marker of astrocyte reactivity, or response to injury. An important point, however, is that increased GFAP was noted by Henriksson and Tjalve (2000) only at high doses that also led to olfactory epithelial damage. In the present study, we found evidence for increased GFAP and TNF-[α] gene expression, among other things, in olfactory bulb tissue; although TNF-[α] message expression was elevated in deeper brain structures, this was not the case for GFAP. TNF-[α] protein expression was also increased in those brain regions where its gene expression was increased.

We did not perform histopathologic assessment of the respiratory or olfactory epithelium and thus do not know if the Mn oxide UFPs caused inflammation at the olfactory mucosa, which could have affected neuronal uptake and translocation. A significant nasal inflammatory response may be unlikely given that there was no inflammation in the lung after 12 days of exposure. Even a 1-day exposure which is far less likely to induce nasal inflammation resulted in significant translocation to the olfactory bulb. Our observation of a decrease in lung lavage fluid [β]-glucuronidase activity after 12 days of exposure is similar to the findings of Bairati et al. (1997), who reported decreases in the activity of this enzyme in plasma from humans exposed occupationally to Mn and lead. These changes were used as a marker of heavy metal exposure. Because the physiologic and pathologic implications of these changes are unclear, histologic evaluations of lung and liver after chronic exposures in animals should be performed in future studies. Our results with $Mn[Cl_2]$ suggest that it is retained to a greater extent than the oxide in olfactory mucosa, thus possibly explaining the damage to those tissues by more soluble forms of Mn delivered at high doses.

The earlier studies of olfactory translocation of nanosized particles of different types (viruses, gold, carbon) together with the present study of ultrafine Mn oxide may imply that all nanosized particles deposited on the olfactory mucosa will translocate to the brain. However, uptake into sensory nerve endings and subsequent translocation is likely to depend not only on size, but also on many other particle characteristics, such as shape, chemistry, surface properties (area, porosity, charge, surface modifications), agglomeration state, solubility, and dose. Although there are no data regarding these parameters for sensory neurons, they affect endo- and transcytosis of nanosized materials. However, our findings may not be directly applied to nanoparticles in general until more data are available on mechanisms controlling neuronal uptake and translocation. We conclude from our studies that the olfactory neuronal pathway represents a significant exposure route of CNS tissue to inhaled solid Mn oxide UFPs. In rats, which are obligatory nose breathers, translocation of inhaled nanosized particles along neurons seems to be a more efficient pathway to the CNS than via the blood circulation across the blood-brain barrier. Given that this neuronal translocation pathway was also demonstrated in nonhuman primates, it is likely to be operative in humans as well.

Toxicogenomics

Toxicogenomics is a new undertaking in the pursuit of human genes relevant to health risk from environmental toxicants and related stress. Such stress spans a range of dire threats in the human condition from ultraviolet to blue light, from particulate diesel exhaust to complex chemical mixtures in the air, from threats in reused aeroplane air to threats in herbal remedies, from threats in the food supply from pesticides to threats from genetically modified organisms. This incredible range of interest is embodied in the Environmental Genome Project, sponsored and maintained in the interest of health by the National Institute of Environmental Health Sciences (NIEHS), and about which much has already been written. The aims of the Environmental Genome Project include a characterisation of important genetic polymorphisms that alter protein function to the benefit or to the determent of individuals upon exposure to environmental stress. The completion of the human genome sequencing and hopefully its successful annotation is a major spur to accelerated progress in the realm of environmental response genes. The program coordinates a robust suite of extramurally funded research grants, Environmental Health Centres and intramural research projects.

An important part of toxicogenomics is the widespread use of high throughput expressed gene analysis or microarray technology. In an absolutely amazing reversal of our traditional abhorrence of "descriptive" surveys, the scientific community has embraced this latest technology as a relatively cheap, reliable, fast, and effective way to protect populations and specific individuals from environmental risk. Why should we be interested in this approach to gene expression studies that uses high-level image processing, high-fidelity polymerase chain reaction (PCR), and clever hybridisation protocols for the purpose of determining gene expression levels by comparison of different RNA populations? In a word scale. The simultaneous examination of a hundred genes is descriptive; the simultaneous examination by comparison and clustering of tens of thousands of genes is a new way to do science.

Basically oligonucleotides or small pieces of cDNA are spotted onto glass slides (or engineered into computer chips), and then RNAs to be compared are hybridised after rabelling with two different fluorochromes: Cy3, which fluoresces red, and Cy5, which fluoresces green. The genes studied may be selected for specific purposes, such as those that are specific to environmental stress, or they may be broad based for gene discovery purposes. Fluorescence appears in the spots (or on the chip) and is detected with image analysis microscopes, usually laser scanning microscopes (or a specialised chip reader). When there is a difference between expression levels in the RNAs being compared, the fluorescent signal with one fluorochrome is greater or less than that of the other. This is detected and quantified.

When the intensity with the two fluorochromes is the same (i.e., there is no difference in the two RNAs), then a merged (third) colour (yellow) appears. The method detects relative changes in differential expression of populations of RNAs, but it may be combined with other techniques such as real-time PCR to obtain quantitative data for selected genes. Depending on the availability of sequence information for the genome of interest and availability of probes from the sequences and the density of individual spots, tens of thousands of expressed genes can be simultaneously examined in a single or a few experiments.

A short parable, though. Suppose one had a large blank wall, and onto this wall one painted tens of thousands of squares, either green, red, or yellow. The squares might be painted in some randomly determined order or by a grand design. Now, suppose your funding agency gave you 100-200 darts rabelled "important new research directions" and instructed you to throw them at the wall. Importantly, you have a map of the squares on the wall, which, like the map on the inside cover of a box of assorted chocolates, identifies all of the squares as genes in a particular genome. You throw the darts and then rush to the wall and look for the struck squares that are green or red and declare them "changed gene expression levels." From those 100 or 200 you may have struck 50 or 60 squares that are "changed gene expression levels." From the 50 or 60 struck by chance you might become genuinely excited by around 10 or 12 of the "genes" that, because of the course of your own research, you believe to be related to whatever question set you on your path to new knowledge. The rest you would declare interesting, important, and necessary to be cherished for a later time to figure out.

While it is possible that gene expression profiling is described by the foregoing parable, early results indicating specificity of array data suggest that there is true inductive value from this process. It is interesting to note that similar uncertainties existed early in the days of gene sequencing. Importantly, the power of that paradigm was realised when hundreds of researchers working independently of each other on seemingly unrelated genes looked at each other's sequences and realised that we had much in common with each other and with the species that we study. That was brought about, of course, by widely accessible, accurate searchable databases of sequence information that we all agreed to use and by dramatic increases in the efficiency of DNA sequencing. Can the same approach work for microarray studies? It seems very probable.

Comparison of protocols, tissues, array production methodology, and myriad other variables will lead to an understanding of which technical details alter the results and which are not important. The NIEHS has committed to creating the tools to do this, and it will take the shape of the National Centre for Toxicogenomics (NCT), which will incorporate the NIEHS Microarray Centre. The NCT will be at the NIEHS and will coordinate both extramural and intramural activities with wide access to the toxicology community. An important part of the NCT will be the establishment and maintenance of a toxicology database for microarray results. The NCT will also support the ToxChip, already developed at the NIEHS. Such a microarray database, GEO (Gene Expression Omnibus), is already up and running at the National Centre for Bioinformatics. There are some problems with the database approach. For example, why would anyone want to enter their data into the program? And how can the quality of the data be assessed before they are released to the public? Here some guidance comes from the other major molecular biology databases.

Data can be deposited and held until publication. Peer-reviewed publication could be taken as a surrogate for the legitimacy of the data, and data could be clustered by experimental approach or by the type of question studied. Techniques in tissue-based arrays could be enhanced through international workshops. These might be part of the National Toxicology Program (NTP), which after all has the most complete set of toxicologic response data in the world, with tissue repositories and sophisticated methods for selection of compounds to study. Results from expressed gene array analyses are very sensitive to the methodology used to establish the initial comparison of RNAs. In some situations it is desirable to determine what genes are expressed in tissues from one physiologic state versus another. In these experiments, the way in which the tissue is obtained will determine the value of the data, insofar as other differences will be extant in the tissues besides those of primary interest. In some instances it may be desirable to determine what genes are expressed in the presence or absence of the expression of another gene of interest. Such an approach is particularly appealing in cascades or networks of interacting genes such as a signal transduction pathway or a developmental hierarchy.

Here we can choose a cell line, transfect in a plasmid that forces high levels of expression of the gene of interest, and then determine what genes change expression levels. The cell line chosen for such an experiment is important. The gene of interest, if transiently expressed, could lead to some subset of genes that respond. In contrast, if we use stable transfection, we might find gene responses that are secondary to the drug selection system. It could be important that there be a recognisable phenotype which varies between the two states examined, although the genes that determine the phenotype might not be related to the gene of interest. It is important to realise that the fluorochromes vary in their stability, so it is important to label the RNAs both ways; that is, a reciprocal hybridisation should be performed in which the Cy3-rabelled RNA is rabelled with Cy5 and the results compared. Low-abundance RNA may be lost, and subtle differences in RNA in states that are compared may be lost due to poor signal-to-noise ratios. However, low-abundance RNA and small changes in RNA level might be very important. The RNA should not be converted to cDNA probes because this may skew results due to preferential amplification of some RNAs relative to others. Even though highly efficient laser scanners and chip readers have been developed and fairly high-density spotted arrays can be managed, the manner in which the array is produced is very important. Poor spot production or misarranged spot reading can cause artifacts in the results. Particularly appealing is the opportunity to profile conditions. For example, if in a difficult diagnostic differential we could establish that a given set of genes was expressed in one disease state but a different set of genes was expressed in another disease state, then determining which genes are expressed in an unknown case could lead to more informed diagnosis. In stratifying patients for treatment, prognosis, and research, such an approach is particularly useful. Here one needs some way of knowing a priori what a given sample is in order to determine the gene expression profile; but if that can be done with confidence, what is the value in the end of knowing the profile? In this circumstance we might hope to use a prospective approach that allows elucidation of the correct diagnosis after the fact by the bahaviour of the disease and with appropriately selected individuals whose RNA can be examined. Moreover, once the analytical method is established, we might expect that it would be rapid and highly reproducible.

A logical extension of the approach of expressed gene profiling is to study physiologic response to environmental stress, in particular toxic exposure, and to determine for any given class of toxicants what genes alter expression in stereotypical ways in a standardised exposure protocol, either of intact test animals or particular cell lines. In this way it seems entirely possible to create a toxicant class-specific profile or fingerprint of expressed genes. The appeal is that then we need only to repeat the exposure protocol for an unknown chemical and declare it dangerous or safe depending on the genes whose expression is altered in the exposure protocol. Obviously the quality of the resulting profiles and their usefulness will depend on the exposure protocol and its appropriateness to the supposed class of the toxicant. Moreover, it will depend on the cluster of toxicants included in a class. If the class of toxicants is too broad, too many genes would be expressed and the genes would not be specific; if the class of toxicants is too narrow, there could be almost as many classes as chemicals to be studied and the application of the profile to the chemical would become hopelessly ad hoc. Indeed microarrays that allow the examination of toxic responses are already available and under close scrutiny.

There are issues that current approaches cannot fully address. In complex signal transduction pathways, the activation of transcription factors is frequently the result of protein-protein interactions and not solely the result of genetic regulation. For example, in the Sonic hedgehog-Patched-GLI pathway the transcription factors GLI1 and

GLI3 are tethered to cytoskeletal elements. In the presence of signalling they are released to be transported to the nucleus as a result of an activation mechanism that is not well worked out, but certainly involves the cleavage of GLI3 and probably phosphorylaton of the binding proteins. Modification of one of these binding proteins, Slimb, may be critical to the activation of the transcription factors. Precisely how this occurs and which other helper proteins are involved in the process will require the same type of high throughput large-scale technology currently being applied to expressed gene sets. Technologies are currently being developed that will allow this, and they no doubt will be important in our study of environmental stress response and attempts to develop toxic signatures for unknown compounds.

Even at the genetic level, not all regulation will occur at the level of RNA transcription and may not be easily detected by subtraction type strategies. For example, the regulation of protein production through translation control is an important aspect of several genetic pathways. In the case of Tra 2 in Caenorhabditis elegans and GLI1 in humans, the 3'UTR in the transcribed RNA contains an element that binds a regulatory protein(s) which leads to shortening of the poly A(+) tail and results in lower levels of protein production. This mechanism and the regulatory binding proteins are conserved from the worm to vertebrates. Clearly, the binding proteins and the elements they bind to are targets of environmental stress. Microarray data cannot help to elucidate regulation that results from subcellular compartmentalisation of molecules, as in the case of p53 or cyclins. Where regulation results from reversible phosphorylation or from ubiquination, the gene expression data will not be helpful.

Arguably the largest, most important environmental challenge facing the American population in the next 25 years is air pollution. Purely as a quantitative matter, the volume of environmental stressors that we expose ourselves to, the sheer surface area of the exposure in our bodies (i.e., the total airway surface area), and the complexity of the chemistry behind it should be compelling enough. When we consider the extraordinary growth of the sources of air pollution in this country, it is hard to imagine a more important environmental health issue. A wide variety of serious pulmonary diseases have not been well characterised, and there is a major need to establish paradigms in health studies that cross over the traditional emphasis on chemistry of exposure versus the study of pathology of lung diseases. This is particularly true of childhood morbidity. In many children's hospitals pulmunary diseases such as asthma and cystic fibrosis represent leading causes of admission. It is important to consider that events in childhood can be precursors of adult chronic disease, greatly increasing the burden on society. Because these events have environmental components, we should establish ways to study pulmonary disease in the context of the complex chemistry of air pollution. This will require sophisticated epidemiology, interventional epidemiology, and robust studies in pulmonary biology. The new genomic information and toolboxes may lead to further progress.

It may be possible to screen biological samples obtained from workers at Superfund sites for the adverse effects of exposure to compounds present in the site. Although the Superfund project has made important advances in public health in the United States, we have a great deal to learn about the effects on workers at these sites. Moreover, we are now beginning to realise the full magnitude of these sites. As we begin to deal with seriously contaminated sites in the Department of Defence and the Department of Energy, the magnitude of true cleanup problems will strike home. We should be prepared to understand the effects on workers performing these cleanups. Once we know a suite of genes that are important in health outcomes at such sites, it may be possible to use real time PCR on the spot as a signature of both exposure and risk.

We must continue to grapple with the question of how the mass of data that will be generated in the next five years can be translated into experiments which will help regulators. One need of toxicogenomics involves the development of sound policy that must be based in good science. The challenge is to first create the good science, then to generate the discussion that informs policy. The standards of analysis, toxicogenomics databases, and public discussion can provide a platform that supports the creation of good science. The challenge will be to figure out how. Many people who are worried about environmental stress wonder if a given exposure in a given setting will pose a threat to an individual's health. What damage occurs if a cloud of diesel exhaust blows in your face? What is the recovery phase from the exposure? What is the health consequence? How frequently does the exposure need to occur before these health effects are permanent? The answers will depend, in part, on the effects of polymorphisms in the response genes, and these are not yet fully known. It will be important to coordinate the efforts of the Environmental Genome Program to find the meaningful polymorphisms with gene expression profiling.

Exposure effects present a longitudinal problem: If the relevant tissue could be obtained from the relevant population, the throughput potential of the new toxicogenomics technologies could allow the dimension of time in accumulated exposure risk to health to be addressed. The challenge is to determine how.

There are large human cohort studies under way nationally in several important areas. For example, the National Cancer Institute and the NIEHS are accumulating long-term data from tens of thousands of agricultural workers with respect to pesticide exposure. Can the tissue and data from the pesticide study be linked to gene expression studies in a meaningful way? The throughput potential of the technologies discussed above should allow this, but we must figure out how to do it. There is a great deal of evidence that points to the effects of environmental stress on infectious disease outcomes. It seems axiomatic that there are important interactions of toxic exposure with pathogens that will effect health outcomes. Can expressed gene array data, gene polymorphism data, or human sequencing lead to identification of overlapping gene pathways in pathogen response?

The completion of the Drosophila genome led to a number of exciting intellectual developments. Among these was a change in biomedical research paradigms toward consortia or engineering models of cooperative effort. This approach has been used in the past, but never, I believe, on such a scale. Another exciting development was the manner in which the genome was annotated and the information that flowed from the annotation. This approach to a massive data set may inform the growing field of toxicogenomics. Can there be an environmental response gene annotation emphasis on the human genome project data?

As we look back over the past 12 or 15 years of genetics, we ask ourselves what we have learned about large efforts and the payoff they can have for important advances and for the growth of the general body of knowledge. What tools have survived the test of time (even though relatively short) and the test of utility? The pioneering efforts of sequence databases, protein structure data, and the accessible tools that allow meaningful use of the data are at the forefront. Rapid emergence of computing tools that do not require intimate knowledge of computer architecture to operate them have allowed important discovery, new knowledge, and, importantly, new experimental designs to flourish. The expressed sequence tag (EST) programs have not only facilitated physical mapping of chromosomes but have allowed innovative use of genome-wide information. The involvement of the National Centre for Bioinformatics in the creation of a database of unique genes in the EST pool [UniGene] has provided highly accessible information and clone sources to investigators worldwide. It should be anticipated that computing resources important to the range of molecular techniques in toxicogenomics (sequencing, polymorphisms, phenotyping, expressed gene microarrays, proteomics) will become available and widely used. The development of tissue arrays should offer a new chance for the NTP to provide a leap forward in our understanding of non-cancer end points of environmental stress.

The NTP has a huge collection of tissue and data from animals exposed to nominated compounds that could be used to determine the longitudinal response of environmentally sensitive genes to exposure. The Environmental Genome Project will also develop such a repository. The Agricultural Health Study will likely develop a repository that could also be used in this way. I hope that the information and tissues can be made broadly available to the scientific community so that they can be fully used to the benefit of public health. As the United States continues to struggle with a toxic legacy and with an ever-increasing burden of environmental stress, especially from air pollution, new methodologies to assess risk to both populations and individuals must be put in place for our protection. How will the scientific community respond to this challenge? In what ways will the genetic revolution of the past several decades contribute to this vital cause? We may hope that as toxicogenomics approaches maturity, the concentration of a suite of tools and the management of large data sets can lead to a point where meaningful science will guide public health policy and minimise personal health risk.

Environmental Geochemistry

The annual increase in world population is approaching 80 million per year (UN 1998), approximately 90% of which is in the poorest countries where one quarter of the population live in 'absolute poverty' as defined by the World Bank (1998). By the middle of the next century world population is projected to be almost double that of the 1950s (UN 1998), although the relative distribution of wealth is predicted to be similar to the present day (World Bank 1998). Hence, it is estimated that by the end of the next century approximately eight out of nine people will live in poor developing countries compared to approximately one in two in 1950 (UN 1998).

Pressures on the surface environment are different in rich industrialised countries compared to those in poorer developing countries. Industrialised countries have frequently had a long history of mining and mineral working, manufacturing, intensive agriculture/ forestry and urbanisation. Europe, for example, which is one of the most densely populated continents, is also one of the most intensely utilised land areas on earth and an increasingly important problem is the need to redevelop derelict and contaminated land. This is becoming more difficult because of tighter legislative and fiscal controls especially the introduction of trigger and action levels for potentially harmful elements and species (PHES), which are being used increasingly as an integral part of the planning and management process. This is despite the need to reuse 'brown field' sites in urban areas rather than extend development into 'green field' areas. Public concern about the environment continues to grow in such countries, and is amplified by the media and results in political pressure. In response, national-international governments and their agencies such as the European Environment Agency (EEA) and the Environmental Protection Agency (EPA) in the USA, are attempting to develop policies, legislation and infrastructure. Substances currently subject to regulation include heavy metals, arsenic and fluoride, radioactive substances and a wide range of potentially toxic organic pollutants. Despite the fact that many of these substances have been subject to regulation, since the 1950s information on safe, trigger and action levels remains controversial. They are often based on limited and/or inadequate information particularly on dose response relationships, natural baselines and exposure analysis.

In many developing countries, population pressure and poverty have adverse inter-related environmental impacts leading for example to widespread deforestation, land degradation and soil erosion from the over-use of marginal farmland and cash cropping. The socio-economic pressures on the environment are frequently compounded by the fragile nature of the surface environment, which as discussed below can be particularly susceptible to degradation and pollution. Each year around six million hectares of the world's agricultural land are lost through soil degradation (UN 1998). Soil is a complex mixture of rock and mineral detritus, organic matter, microbes (bacteria, yeast and fungi) and invertebrates. The removal of protective plant cover, overgrazing, overcropping leading to the depletion of essential nutrients, or salination (in some cases as a result of irrigation and overpumping of aquifers) can cause soil organisms to die and soil to turn to inert mineral dust (desertification). Artificial irrigation schemes based on the increased use of groundwater in such environments can greatly increase levels of F or Se in the surface environment or As as in the recent example of Bangladesh.

Historically, mining has also caused contamination, including cyanide, mercury and arsenic releases from gold mining, and radioactive element pollution from tin and uranium mining. Although large international mining companies now generally work to high environmental standards, mineral working by uncontrolled and disorganised groups (informal mining-especially for gold-continues to cause environmental problems in developing countries as in the recent incidents of mercury pollution in Brazil. More recently intensive industrialisation is having other serious adverse environmental impacts for example the largescale burning of low quality coal in India and China. Some of these coals have exceptionally high contents of radioactive and potentially toxic chemical elements which are released to the environment together with CO, and SO, and NO, which are increasing problems associated with the accumulation of greenhouse gases in the atmosphere and acid deposition. In addition, low quality coals can have a very high ash content which results in problems of residual ash disposal and associated heavy metal leaching.

Economic and population pressures on rural economies cause migration, especially to urban centres. In developing countries the growth of very large cities (with populations greater than five million) is of particular concern; on current trends the population of some eight to ten cities will reach 15-25 million by the end of the century (UN 1991). Urbanisation in turn leads to further adverse environmental effects, such as contamination of soils, surface water and aquifers through poor sanitation. Urban pollution and the accumulation of potentially harmful wastes are particularly intense in some developing countries. Pollution of air, water and soil from vehicles, power plants and factories is at its most extreme in cities, with increased incidence of childhood diseases associated with heavy metal poisoning.

In addition to the scientific and technical aspects of the problems, socio-economic factors are important, Communication on environmental issues can be particularly difficult where society is opposed to birth control and where there is a perception that 'the West' in its concern for the environment is trying to limit economic growth. Increasingly, problems in regions such as Sub-Saharan Africa appear overwhelming.

Historically the response of the international community has been mainly one of reacting to the most acute problems as they arise. However, this policy is increasingly seen to be failing particularly in providing fundamental long-term solutions.

An alternative approach proposed here involves developing countries and industrialised countries working together to document the problems of the global environment with methodologies and information made easily accessible over the world wide web. The preparation of a global geochemical baseline is regarded as an essential first step for assessing and monitoring the state of the earth's land surface. Initially the aim is to map the distribution of chemical elements particularly those that are essential or potentially harmful to life including the natural and anthropogenic radioactive elements. In the future, the distribution of potentially harmful substances especially organic species could be included. Such a strategic approach has the potential to help in the development of policies for the more sustainable use and development of land into this millennium.

Essential and Potentially Harmful Chemical Elements

Two main groups of chemical elements are of particular importance for health; essential and potentially harmful chemical elements. Those identified as essential to animal and plant life include K, Na, Ca, P, Cl -, SO, and NO, as well as the first row transition elements Fe, Mn, Ni, Cu, V, Zn, Co and Cr (Mertz 1986) and Mo, Sn, Se, and I. Some of the disorders associated with deficiency of these elements are given in Mertz 1986. Boron has not yet been shown to be necessary for animals, although it is essential for higher plants and has been shown to be toxic in man. By contrast potentially harmful elements (PHEs), known to have adverse physiological significance at relatively low levels, include As, Ag, Be, Cd, Pb, Hg possibly the rare earth elements Ce and Gd and some of the daughter products of U. Aluminium can also have adverse physiological effects in trace amounts in animals and particularly in fish and plants and has been implicated in clinical conditions in man including in some types of neuorological diseases. All trace elements are toxic if ingested or inhaled at sufficiently high levels for long enough periods of time. Selenium, F and Mo are examples of elements which show a relatively narrow concentration range (of the order of a few [tg - ‘) between deficiency and excess (toxic) levels. The difficulties of diagnosing disease, and particularly subclinical conditions, in animals and man related to trace element deficiencies or excesses are discussed by Mills (1996). Except in some specific cases such as arsenic or mercury toxicity and for example iodine deficiency disorders (IDD), symptoms may be non-specific and diagnosis, based on tissue or blood sampling, costly.

The concentration of essential trace elements such as Se in crops (rice, corn, soybean) was shown by Levander (1986) to correlate closely with the concentrations in the soil in which they are grown, although the relationship has recently been shown to be more complex in the case of Se. Presently much effort is focused on groundwater as the principal pathway for many PHEs to man.

However, the situation is considerably more complex with soil, aerosols, dust and diet being more significant for most elements with the exception of F in contributing to the body burden of both essential and PHEs. For example, recent work by Williams et al. (1999) using physiological based extraction techniques (PBETs) and risk assessment has shown that soil is likely to be an equally significant exposure pathway for as in many developing countries. Significant soil contamination can occur throughout harvesting and food preparation processes and in some countries the direct consumption of soil especially by pregnant women (geophagy) is encouraged.

Geochemistry can identify not only the total amount of each element in soil, dust or water for example, but also the amount which is bioavailable. This is more important than the total amount but is less easy to establish and may be contentious. For example Al, which is the commonest metal in the Earth's surface layer, occurs in both inert and bioavailable forms, and its potential toxicity depends on its chemical form or speciation. It can occur in a large number of dissolved aqueous species, especially in conditions of low or high pH, or in association colloids with organic carbon or silica which limit its bioavailability and, therefore, toxicity. The toxicity of As and Sb also depends on their chemical form. They are most toxic in the M - 3 gaseous state with decreasing toxicity in the sequence M3*> Ms+> methyl As/Sb.

In higher animals, including man, assimilation occurs by ingestion of nutrients and contaminants by the inadvertent or advertent ingestion of soil, food crops, animal tissues, drinking water, absorption through the skin, and by respiration. Ideally geochemical data on soil and dust, stream sediment and ground and surface water should be available to identify, assess and model potential toxicity or deficiency situations.

Chemical Speciation

The speciation of chemical elements affects their distribution, mobility and toxicity. This has been known for a long time in the case of the common anions (HCO -, NO-, SO'-) and surface environment more recently for a wide range of chemical elements in relating geoche cal data to health was first established by agricultural scient.

The Importance of Chemical

Recently con in relating geochemical developments have been made in the prediction of chemical data to health was first established by agricultural scientists speciation by modelling thermodynamics. Recently considerable and by experiments have been made in the prediction of chemical determination by modelling thermodynamics. Biochemical processes, induced by microbial, plant and by experimental activity, are also important controls on both speciations. Biochemical processes, induced by microbial, plant and animal actively poorly understood until recently also important controls on both speciation and mobility, Some of the most important controls, particularly understood until recently (Ehrlich element speciation and mobility, include hydrogen & Goodman activity

Some ox potential (Eh), the most important controls, particularly on trace properties of solids, the abundance and speciation of potential ligand mobility, include hydrogen ion activity (pH), redox por cations and al (Eh), temperature, surface presence or absence of dissolved solids, the and/or particulate organic matter, and speciation of potential activity. Two major cations and anions, the presence or absence of dissolved and/or particulate organic matter, and biological activity. Two of the most important factors directly controlling mobility and solubility are Eh and pH. The solution chemistry of an element is affected profoundly by changes in oxidation state, while dissolution reactions, including hydrolysis, inorganic complexation, complexation with smaller organic anions (such as oxalate) and sorption/desorption, are all pH controlled. Under conditions of high pH, anions and oxyanions (such as those of Te, Se, Mo, U, As, P and B) are more mobile and most cations (such as those of Cu, Pb, Hg and Cd) are less mobile, while at low pH the reverse is generally true. Where humic substances or biological by-products are present, however, stable organometallic complexes (which may behave as anions) are formed, increasing trace-element mobility. The kinetics of inter-species interactions also act as important controls on speciation, particularly where natural systems are disturbed by the influence of man. Geochemical interaction between the geosphere, hydrosphere and the biosphere depends partly on sorption processes and partly on chemical speciation. Some elements, for example Al, Ti and Cr, are relatively poorly assimilated by plants, although others such as Cd, Se, Mo and Co, can readily cross the soil-plant barrier and enter the food cycle. In soils, sorption of elements by clay minerals and organic material is the predominant fixing mechanism, with soil pH controlling sorption processes and metal solubility/ bioavailability. Many metals are more soluble in the low pH conditions induced by natural organic acids and root exudates.

Speciation studies are of particular importance in areas affected by land degradation, deforestation or pollution caused by mining, industrial activity or urbanisation, and they can be used to optimise amelioration strategies and improve management practices. A knowledge of the factors controlling speciation in different environments can be used to predict the potential for absorption of PHEs.

Speciation and lability both in the natural environment and in the gastro-intestinal tract, exert a major influence on the uptake and assimilation of trace nutrients and PHEs. Several workers have recently developed physiologically based extraction tests (PBETS) that attempt to predict the bioaccessability (the total fraction of an element that is available for absorption during transit through the small intestine) of a given trace element. Such tests have significant advantages over conventional sequential leach tests because they attempt to reproduce physiological factors, such as stomach emptying time and the ratio of solid to digestive fluids as well as the pH and Eh of upper gastrointestinal tract solutions, in a particular target group, such as young children. In the case of As the values for bioavailability in the upper intestine (neutral pH) are consistently significantly greater than those estimated from simple acid stomach solutions. Although, these tests need further calibration and validation they represent an important step forward in understanding exposure pathways for both potentially toxic trace and essential trace elements.

Behaviour of Chemical Elements in Different Surface Environments

Both the total quantities of chemical elements and their speciation are likely to vary between temperate and tropical terranes. In cool temperate to Mediterranean climatic regions soils tend to be of moderate thickness and comprise mainly podzols and brown earths developed on landscapes dating mainly from the Cretaceous to the Pleistocene at high latitudes. Chemical conditions including the electrical conductivity and activity of major cations and anions are also moderate while relatively high levels of organic matter tend to complex and buffer levels of PHEs and essential elements such as I and the other halogens.

At lower latitudes and particularly in continental areas of low relief, the regolith may be an expression of the cumulative effects of subaerial weathering during many millions of years. Some soils have been deeply leached under various climatic conditions, including long periods of high rainfall and temperature to which they have been repeatedly subjected from Tertiary times or even as far back as the late Proterozoic. Tropical and semi-tropical soils, partly because of their age and the number of weathering cycles to which they have been subjected, tend to be poorer, thinner and more fragile than those in temperate regions. This is particularly the case in shield areas of Africa and Latin America where soils have developed on crystalline basement and are generally older. In contrast, those of SE Asia developed on alluvium or volcanic rocks are generally younger and more fertile.

In tropical countries laterites, which have been forming continuously for at least 100 million years (since the Jurassic) in parts of Africa, India, South America and Asia, are common. Weathering profiles may extend to depths of over 150 m, with weathering fronts at great depth, so that the fresh supply of chemical elements from rocks into the biosphere is limited. Laterites, which include Fe- and Al-rich tropical weathering products such as bauxites, ferricretes and Fe- and Al-duricrusts, have markedly homogeneous mineralogical and chemical compositions which usually bear little relationship to underlying bedrock. They have well-defined chemical profiles, but, depending on the climatic and tectonic history, the land surface may be highly complex with the profiles truncated or buried by later detritus.

In such conditions, intense oxidation, which generally extends to the weathering front, results in a lack of organic matter (which is stored in the biota rather than soils) and of other major elements such as phosphorus, nitrogen and potassium, and also in the formation of stable insoluble minerals such as clays (especially kaolinite) which markedly increase the Al/Si ratio. Generally, such soils are deficient in soluble cations (Na+ and Ca") and anions (CI -, PO' - and NOt), but have high levels of resistant oxides of Fe, Al, Ti, Ce and Mn. They are typically kaolinitic or ferralitic, with local calcretes or silcretes in areas of inland drainage affected by evapouration and a fluctuating water table. Hardening of aluminium and iron oxide to form cuirasses, which armour the otherwise friable surface, is common in tropical terrains. As natural vegetation is cleared from such environments, essential nutrients are further leached and the top soil may be completely eroded. Evidence exists that some deserts were formerly subjected to extreme leaching under tropical conditions during the Cretaceous and Tertiary. Most desert soils are typically lithosols, the important constituents of which are resistate detrital quartz and accessory minerals such as zircon and secondary clay minerals. Organic material in desert soils is minimal and surface waters are highly oxidising, with variable (frequently high) pH values. They generally have higher salinity than those in other environments, with chloride rather than bicarbonate as the main anionic species.

Chemical elements tend to undergo more pronounced separation in tropical environments than in temperate regions, although this depends on their chemical mobility and on the nature of the local environment. Hence areas with the potential for deficiency or toxicity conditions are likely to be much more common in tropical countries than in temperate regions.

The most deeply leached environments retain only the most inert elements such as Fe, Mn, Zr, Hf and the rare earth elements (including Ce), either as secondary or primary oxides. On the other hand the most mobile elements, such as the alkalis, alkaline earth elements, the halogens and elements mobile in conditions of high pH including anions and oxyanions, such as those of B, Se, Mo, V and U can accumulate in and environments. These include inland drainage systems and more locally the base of weathering profiles. The variable oxidation states of the first row transition elements Co, Cr, Cu and Zn also favour removal from solution by ion exchange, precipitation and surface sorption. The lack of organic matter as a result of intense oxidation can mean that total levels of such elements in the regolith may be high, but they may be bound on Fe-Mn oxides and thus their bioavailable levels may be exceptionally low, resulting in potential deficiency for plants and animals, especially in deeply oxidised surface environments. The controls on the speciation of As, Cd and Pb are broadly similar to those of the first row transition metals, though the oxidation states of these elements, with the exception of as, exhibit less control on mobility.

Geological Survey Organisations and the Sustainability of the Earth's Land Surface

In most countries geological survey organisations (GSOs) provide strategic information on the Earth's land surface, and in the case of industrialised countries they tend to be one of the main instruments for providing international aid and development assistance in this area. Until the 1950s GSOs were principally concerned with the preparation of geological maps as a basis for economic development. More detailed or generic studies tended to be concerned with such studies as the origin of granite batholiths or the structural history of particular orogenic belts. Between the 1950s and 1980s there was an increased use of geophysical and geochemical survey methods partly to provide additional information with which to prepare and interpret geological maps and partly to identify new, especially metalliferous mineral resources. In the case of geochemistry relatively little emphasis was placed on standardising methods which were increasingly divergent in relation to different exploration targets and different terrains.

More recently as concern about the environment has increased more emphasis has been placed by national governments and international aid agencies and development banks on studies concerned with such issues as the quality of water resources; waste and pollution related to historical mining, industrialisation or urbanisation; and, in developing countries, soil degradation as a result of salination and deforestation. Some examples of the most recent studies carried out by the BGS at home and internationally. Such studies make an important contribution to applied environmental research. In developing countries they improve understanding of the chemical processes operating at the land surface in tropical, equatorial and desert terrains and typically they now involve scientists from both donor and recipient countries. Nevertheless much more needs to be done if any substantial impact is to be made on global environment problems.

Limitations of the type of studies being carried out include the tendency to investigate only specific media; and groundwater studies in particular tend to be carried out in isolation from those of surface water, stream sediment or soil. There is also a tendency to study one or two chemical elements in detail, an approach that fails to provide a complete picture of interaction between chemical species in the surface environment as a whole. Methods of sampling, analysis and data interpretation often vary greatly. Moreover, such audits are generally published in the journals of learned societies and are communicated mainly to other experts in the same field with relatively little impact on science or· society as a whole.

Increasingly fully integrated digital multi-media multielement geochemical data are available in industrialised countries and some examples of their application to environmental studies are discussed below with particular reference to Britain and the USA. When linked to process and model studies such data provide a powerful means of documenting problems, providing solutions and subsequently assessing their effectiveness. Developing countries are increasingly recognising that geochemical mapping should be given a high priority within their national programmes.

Some Applications of Modern Digital Geochemical Databases

Geochemical survey methods based on rapid low-cost analysis easily collected surface materials were developed initially for exploration purposes. In Britain, the USA and elsewhere stream-sediment sampling has been the preferred option. Geochemicat maps particularly those prepared to modern standards are still highly effective in identifying metalliferous mineral prospects with applications in exploration and in land use planning where the aim is to avoid sterilising potentially valuable resources. Modem methods are considerably more sophisticated than those developed for prospecting earlier this century however. An increased range of sample media is used including stream sediments, soil, and surface and shallow groundwater samples ideally accompanied by airborne radiometric cover to map the distribution of the naturally occurring radioactive series (U, Th and K) and anthropogenic isotopes such as Cs17 and Sr". Sampling is increasingly carried out to rigorous internationally agreed standards, with analysis based on state of the art high precision methods. Sophisticated methods of data processing are used including the use of thermodynamic models and speciation codes and strict quality control and quality assurance procedures are maintained throughout. The principal features of the G-Base programme in Britain are given. In the USA the National Geochemical Database held by the US Geological Survey contains approximately 2 million records of chemical analyses for a range of media including rocks, stream sediments, soils and vegetation. Approximately 65% of the US land surface has been covered at varying densities over the past 30 years for the evaluation of mineral resources. The data are now being used for environmental studies.

The applications of such data now extend well beyond those of metalliferous mineral exploration or resource assessment and improved interpretation of the geological map for which the surveys were initiated. They are used for a range of environmental applications with particular emphasis on the data on the chemistry of soil and surface and shallow groundwater systems. Such data can be used to show the distribution of total element concentrations in particular media or in combinations using simple colour addition methods using GIS. They can also be used with speciation diagrams based on pure ideal systems or with ionic dominance plots or critical load models, for example to provide process information and decision support systems at the national to regional scale. Three selected examples of such applications are described below.

Identification of Contaminated Land in Britain for Planning Purposes

Remediation of contaminated land is expensive, and planning is one of the most effective ways of minimising cost. For example, if the distribution of contaminated land is known, infrastructure such as roads, railways, factories, shops, offices and leisure complexes can be planned to contain the contamination (provided that there is no impact on vulnerable aquifers) leaving 'clean' areas for family housing, agriculture and horticulture.

In the UK a study carried out by Appleton et al. (1998) using the new G-Base regional geochemical dataset where it is available supplemented by data from the Wolfson Geochemical Atlas, identified all of the main areas of Cu-Pb-Zn-Cd-As contamination. These were shown to be related principally to historical metalliferous mineral working, urban and industrialised centres as well as areas with high natural levels of PHEs reflecting the presence of particular rock formations. The final maps which were prepared on behalf of the Department of the Environment are now used as part of the planning process to identify areas where specialist guidance is required at the national to regional scale. Such maps enable values for pollutants over contaminated land to be seen clearly in the context of natural geochemical variation that is frequently as great or greater than that related to man-made contamination. Since the data are held in digital form they can be reprocessed readily as trigger and action levels are changed by regulatory authorities.

Studies of water quality in Wales and western England Studies of water quality in Wales based on the G-Base dataset, have shown the potential for studying a range of environmental problems, ranging from heavy metal contamination related to historical mining and catchment acidification to problems associated with nutrient enrichment that may lead to the development of algal blooms and ultimately freshwater anoxia.

The data provide excellent information on pH, conductivity and the total concentrations of about 50 determinands. They can also be reprocessed to provide fundamental process information. For example the data on major ions can be shown as a percentage of total negative or positive charge ionic dominance as described in Hem (1985). They can also be represented as three colour component images based on the diagrams developed by Piper (1944) to show variations in the relative proportions of Na+, K+, Mg‘ and Ca”; and HCO, SOQ-, NOr- and Cl - respectively. Such maps clearly identify areas of intensive agriculture with high NO; concentrations; areas of high acid precipitation with low buffering capacity (low Ca” and low Mg2+ with high SO’-) and areas of acid mine drainage with high S042 and frequently Cl- in areas of acid mine drainage with high , and former coal working for example. quently Cl- in areas of former coal load maps can be prepared using models and displayed in a similar way as well as maps can be prepared using models and degree of calcite in a similar way as well as maps solubility based owing their solubility degree of calcite or gypsum surface waters studied have low ion their strength (ility diagrams. quartile surface waters studied have of ionic strength (I) from 0.001 to 0.1007 M, with range of 0.146). Hence, geochemical speciation c strength (1) from 0.00 1 to 0. 1007 M, such as max of 0.146). Hence, geochemical speciation can be applied, for example in modelling the bioavailability of PHEs such as PHREEQ and WATEQ can be applied, for example modelling the bioavailability of PHEs such et as Pb.

The wide range of determinands and the high sample density used which is based on low-order streams that are closely related to catchment geochemistry provides a better means of assessing overall environmental quality than data normally collected by regulatory agencies for monitoring purposes.

Global Geochemical Baselines

Extending modern systematic geochemical databases such as those prepared by the BGS and USGS over the Earth's land surface clearly provide one of the most effective methods of addressing environmental concerns worldwide. However, presently available data are incomplete and frequently inconsistent between different countries, while many older datasets which were collected mainly for prospecting purposes, especially in developing countries, do not meet even the basic requirement for establishing national environmental baselines. Moreover, while there are exceptions, in the past different national and international agencies have carried out different aspects of geochemical surveys using different methodologies. Hence geological surveys have traditionally provided chemical data on rocks and stream sediments; soil surveys chemical data on soils and hydrological survey organisations data on the chemistry of ground and surface water. It is difficult if not impossible to interact such datasets using GIS or to use them as a basis for studying processes operating in the Earth's surface environment as a whole.

The concept of internationally standardised geochemical mapping procedures originated in the 1970s under the auspices of the International Atomic Energy Agency, specifically for uranium. Successful application of the concept led, in 1984, to discussions on the need to have standardised data for all elements. In 1988 formal support was obtained from the IUGS/UNESCO International Geological Correlation Programme (IGCP) for what has become the Global Geochemical Baselines project.

Initially a comprehensive review of sampling, analytical and data processing methods in use was carried out. Based on these findings and it was recommended that a framework be developed for the adoption of standardised methods and reference materials for geochemical mapping of the global-regional scale. It was agreed that this should eventually lead to the development of a common primary global geochemical baseline database that would be multi-media and multideterminand. The recommended sampling, analytical and data processing methods are described in Darnley et al. (1995); expanding upon this information, a detailed sampling manual for temperate regions has been prepared by Salminen et al. (1998). Wide spaced sampling is required over the entire land surface, including regions already surveyed to enable available datasets to be levelled as well as regions where there is a low probability of any geochemical mapping being carried out in the near future. The samples collected will also serve as standard reference materials.

The GRN of samples has been completed for China and parts of Russia and work is on schedule to complete the work over 27 European countries by the year 2000. Australia is carrying out an airborne gamma ray survey of the whole country as a component of the project. A considerable amount of work on developing a geochemical baseline is also being carried out in Southern Africa, Colombia, Brazil, Korea and many other countries. Also, geological survey organisations preparing geochemical maps in developing countries are increasingly using the standard methods. Much more financial assistance is required however to extend the network especially over developing countries. As indicated by Darnley et al. (1995) the total amount of money required for such an important study of the Earth is no greater than the launch preparation costs for one space shuttle mission.

Land degradation and pollution related to population pressure, economic development and in developing countries, poverty are increasingly affecting the sustainability of the earth's land surface and its life support systems. The problem is most acute in tropical, equatorial and desert regions where the surface environment is particularly fragile because of its long history of intense chemical weathering over geological timescales.

Specific process studies of particular pollutants, pollution incidents or areas of land degradation and desertification provide valuable generic information but are unlikely in themselves to have a significant impact on the management of global environmental change. On the other hand, preparation of a systematic multi-media (surface and groundwater, soil and stream sediment) multi-determinand global geochemical database would make a major contribution to improving environmental quality worldwide. Such a programme could involve all countries with the communication of methods, data and information over the world wide web. This would provide the best method of technology transfer; and of documenting and communicating information on the environment of particular regions and countries in a global context as a basis for sustainability. It would also provide a sound basis for targeting scarce resources and developing strategies for remediation and amelioration. A global database would also provide a baseline against which future change could be measured. It is recommended that governments, international agencies and development banks increase the proportion of resources spent on preparing strategic digital databases for the Earth and its environment as a basis for developing policy and planning for sustainable land use into the next millennium. A global environmental geochemical baseline should be one of the highest priorities for such a programme.

Bibliography

Adamson, A. W; Gast, A. P.: *Physical Chemistry of Surfaces*, Wiley-Interscience; New York, 1997.

Adkins, C.J.: *Equilibrium Thermodynamics,* Cambridge: Cambridge University Press, 1983.

Arnaudov, M.: *Application of Infrared Spectroscopy in Chemistry,* Nauka i Izkustvo, Sofia, 1981..

Arora, G D : *Liquid Crystals and Polymers*, Sarup and Sons, Delhi, 2005.

Astarita, G., D. W. Savage, and A. Bisio: *Gas Treating with Chemical Solvents,* Wiley-Interscience, New York, 1983.

Attwood, A. T.; Florence, A. T.: *Surfactant Systems, Their Chemistry, Pharmacy, and Biology*, Chapman and Hall, London, 1983.

Balescu, Radu: *Statistical Dynamics: Matter Out of Equilibrium*, Imperial College Press, London, 1997.

Balkwill, Dr. Fran: *DNA Is Here to Stay,* Carolrhoda Books, Inc., Minneapolis, 1993.

Benvenuto, E.: *An Introduction to the History of Structural Mechanics*, Springer-Verlag, New York, 1991.

Binay Kumar and R.P. Tandon: *Advances in Technologically Important Crystals*, Macmillan Publishers India, Delhi, 2007.

Bohren, C.F., and B. Albrecht: *Atmospheric Thermodynamics*. Oxford University Press, 1998.

Bond, A. Curtis: *Modern Polarographic Methods in Analytical Chemistry*, M. Dekker, New York, 1991.

Bontchev, P.R., *Complexation and Catalytic Activity*, Nauka i Izkustvo, Sofia, 1972.

Braden, J.F.: *Mechanical Fastening of Plastics - An Engineering Handbook,* Marcel Dekker, New York, 1984.

Campbell, I. M.: *Introduction to Synthetic Polymers, USA*: Oxford University Press, 2002.

Clark, M. M.: *Transport Modeling for Environmental Engineers and Scientists,* Wiley, New York, 1996.

Cotton, C.A.: *Volcanoes as Landscape Forms*, Hafner Publishing. New York, 1999.

Coulson C. A.: *Supplementary Tables of Molecular Orbital Calculations*, Pergamon Press, Oxford, 1965.

Crompton, T. K.: *Analysis of Polymers*: America: Pergamon Press, 2005.

David Baraff: *Dynamic Simulation of Non-Penetrating Rigid Bodies*, Cornell University, Ithaca, NY, 1992.

Donald A. McQuarrie: *Statistical Mechanics,* Harper and Row, UK, 1976.

Dreybrodt, Wolfgang: *Processes in Karst Systems Physics, Chemistry and Geology,* Springer-Verlag, New York, 1988.

Duxbury, G.: *Infrared Vibration-Rotation Spectroscopy, From Free Radicals to the Infrared Sky*, Wiley, NJ, 1999.

Esumi, K.; Ueno, M.: *Structure-Performance Relationships in Surfactants*, Marcel Dekker, New York, 2003.

Eugene Allgower and Stefan Gnutzmann: *Simplicial Pivoting for Mesh Generation of Implicitly Defined Surfaces*, Computer-Aided Geometric Design (Netherlands), 1991.

Folk, R.L.: *The Petrology of Sedimentary Rocks*, Hemphill Publishing Co., Austin, 2001.

Fritz Helmet: *Diffraction in Crystals*, Ivy Publishing House, Delhi, 2011.

Gurugnanam, B. : *Essentials of Hydrogeology*, New India Pub, Delhi, 2009.

Halliwell, J.J.: *Physical Origins of Time Asymmetry*. Cambridge, 1994.

Harlyn Baker, H. : *Building Surfaces of Evolution: The Weaving Wall*, International Journal of Computer Vision, 1989.

Herbert, N.: *Quantum Reality: Beyond the New Physics*, New York: Doubleday, 2003.

Hiemenz, P. C.; Rajagopalan, R.: *Principles of Colloid and Surface Chemistry*, Marcel Dekker, New York, 1997.

Higuchi, T. : *Solubility Behavior ofOrganic Compounds*, Wiley-Interscience, New York, 1990.

Holmberg, K., : *Novel Surfactants: Preparation, Applications, and Biodegradability*, Marcel Dekker, New York, 2003.

Holmgren, D. A., Moody, J. D. : *The Structural Setting for Giant oil and Gas Fields*, Tokyo, 1975.

Israelachvili, J.: *Intermolecular and Surface Forces: With Applications to Colloidal and Biological Systems*, Academic Press, San Diego, CA, 1992.

Jackson, J.D.: *Classical Electrodynamics*, John Wiley and Sons, New York, 1975.

James F. Shackelford: *Introduction to Materials Science for Engineers*. Prentice Hall, 2009.

Jaynes, E.T.: *Probability Theory: The Logic of Science*, Cambridge University Press, Cambridge, 2003.

Keizer, J.: *Statistical Thermodynamics of Nonequilibrium Processes*, Springer-Verlag, New York, 1987.

Keller, G.: *Separations: New Directions for an Old Field*, AIChE, New York, 1987.

Khalili, A.: *Quantum: A Guide for the Perplexed*, London: Weidenfeld & Nicholson, 2007.

Konopinski, E.J.: *Electromagnetic Fields and Relativistic Particles*, McGraw-Hill, New York, 1981.

Kurt Georg: *Numerical Continuation Methods, An Introduction*, Springer-Verlag, New York, 1990.

Lehninger, Albert: *Principles of Biochemistry*, Worth Publishers, 1993.

Lifschitz, E. M.: *Fluid Mechanics*, Pergamon Press, North Holland, 1987.

Lindon, John C.: *Encyclopedia of Spectroscopy and Specrometry*, Academic Press, Uk, 1999.

Majumdar P.K. Basu: *Law of Mines and Minerals*, Universal Law, Delhi, 2011.

Marjorie C. Caserio: *Basic Principles of Organic Chemistry*, W. A. Benjamin, Inc., 1964.

Mary Lambert: *Crystal Energy*, Cico Books, Delhi, 2010.

Miller, J. M.: *Separation Methods in Chemical Analysis*, John Wiley and Sons, New York, 1975.

Muynck, W. M.: *Foundations of Quantum Mechanics, an Empiricist Approach*, Dordrecht: Kluwer Academic Publishers, 2006.

Neil Gershenfeld: *The Physics of Information Technology*, Cambridge Univ. Press, Cambridge, 2000.

Null, H. R.: *Handbook of Separation Processes*, Wiley–Interscience, New York, 1986.

Olagoke Olabisi: *Handbook of Thermoplastics,* Marcel Dekker, New York, 1997.

Pramod O. Alexander: *A Handbook of Minerals, Crystals, Rocks and Ores*, New India Pub Agency, Delhi, 2009.

Prutton, Martin: 1994 *Introduction to Surface Physics*. Oxford University Press. 1994.

Roland, O.: *Understanding Quantum Mechanics*, Princeton: Princeton University Press, 1998.

Rosen, M. J.: *Surfactants and Interfacial Phenomena,* Wiley, Hoboken, NJ, 2004.

Ruthven, D. M.: *Principles of Adsorption and Adsorption Processes,* John Wiley and Sons, New York, 1984)

Schwuger, M. J.: *Detergents and the Environment*, Marcel Dekker, New York, 1996.

Seidel, A.: *Kirk-Othmer Encyclopedia of Chemical Technology*, Wiley-Interscience, New York, 2004.

Simeonov, V.: *Information Aspects in Chemical Analysis*, Obrazovatelni Tekhnologii, Sofia, 1999.

Sutton, A.P.: *Electronic Structure of Materials*. Oxford Science Publications, 1993.

Tanford, C.: *The Hydrophobic Effect. Formation of Micelles and Biological Membranes*, J. Wiley, New York, 1980.

Tej Verma: *Promoting Community Nutrition Through Participatory Monitoring and Evaluation*, Advance, 2001.

Thompson, L. M.: *Theoretical Hydrodynamics*, Macmillan, India, 1968.

Treybal, R. E.: *Mass-Transfer Operations*, McGraw-Hill, New York, 1980.

Wahab, M. A. : *Essentials of Crystallography*, Narosa, Delhi, 2009.

Walas, S. M.: *Phase Equilibria in Chemical Engineering,* Butterworths, Reading, MA, 1985.

Whitmore, G : *Chemical Crystallography And Lipid Crystals*, Sarup Book, Delhi, 2009.

Wolf, F. J.: *Separation Methods in Organic Chemistry and Biochemistry,*Academic Press, New York, 1969.

Index

□□□